# Microbes at Biomedical Interfaces 2021

Topical Conference at the 2021 AIChE Annual Meeting

Boston, Massachusetts, USA and Online
7 - 11 November and 15 - 19 November 2021

ISBN: 978-1-7138-5692-4

**Printed from e-media with permission by:**

Curran Associates, Inc.
57 Morehouse Lane
Red Hook, NY 12571

**Some format issues inherent in the e-media version may also appear in this print version.**



Printed with permission by Curran Associates, Inc. (2022)

For permission requests, please contact AIChE
at the address below.

AIChE
120 Wall Street, FL 23
New York, NY 10005-4020

Phone: (800) 242-4363
Fax: (203) 775-5177

www.aiche.org

**Additional copies of this publication are available from:**

Curran Associates, Inc.
57 Morehouse Lane
Red Hook, NY 12571 USA
Phone: 845-758-0400
Fax: 845-758-2633
Email: curran@proceedings.com
Web: www.proceedings.com

# TABLE OF CONTENTS

# 636261 Vapor-Deposited Biointerfaces for Fouling Reduction and Regulated Microbial Behaviors

**Abstract:**

Vapor deposition techniques provide an effective way to control the material properties at a biointerface, where synthetic materials and living organisms meet and interact. The interactions between functional polymer surfaces and a ubiquitous biological system – bacteria, are of great interest in applications ranging from biomaterials and drug delivery to fouling control in membrane separation processes. Recent advances in the initiated chemical vapor deposition (iCVD) of functional polymer thin films have enabled nanoscale control of the synthesis and processing of novel antifouling coatings and revealed the complex effect of polymeric surfaces that is beyond the conventional "killing" (antimicrobial) or "repelling" (antifouling) functions. This talk will highlight the recent progress in (1) the distinct design principles of antifouling polymeric coatings that are effective at the air-liquid-solid triple interface, which differ from the conventional liquid-solid interface presented by submerged surfaces; and (2) an example of insoluble material to exert control over microbial biochemical activities, which led to increased biofilm formation while suppressing virulence, hinting at the potential of polymeric materials to program microbial functions. We seek to underscore the importance of understanding detailed microbe-material interactions and to provide an outlook on extending the material-bacteria interactions beyond "kill or repel" towards signaling and control.

# 636262 Development of Biomimetic Nanostructured Surfaces to Combat Bacterial Adhesion and Growth without Antibiotic Treatments

**Abstract:**

The health and safety concerns caused by bacterial adhesion on surfaces, along with the fast evolution of bacteria with antimicrobial resistance, increase the demand for developing next-generation antibacterial surfaces. Inspired by bactericidal insect wing surfaces, we have developed the biomimetic protrusive nanostructures on silicon and polymer thin films to mechanically kill bacteria upon adhesion without the use of antibiotic treatments and antifouling coatings. We have investigated how bacteria can be killed in mechanical and physical manners by the surface nanostructure as a function of its geometry and stiffness, also studied the bactericidal efficacy against different types of pathogenic strains. In this talk, I will discuss the correlation between nanopillar density and bactericidal efficacy against E. coli as well as the effect of the stiffness of polymer thin films with the optimized nanostructure to kill different types of bacterial strains. The fundamental insight from this work into the role of structural and mechanical properties of nature-inspired nanostructure in bacterial adhesion will offer practical applications for designing diverse polymer thin films such as biomedical implants, food packaging, and public screen touchpads.

# 636191 Rapid Adhesion Testing of Biological Films By Laser-Induced Stress Wave Loading

**Abstract:**

Biofilm formation is a significant problem within the American healthcare system, accounting for 17 million infections, 550,000 deaths, and an estimated cost in the billions of dollars annually. Infections from biofilm formation are prolific in implantology, accounting for half a million cases annually. While there have been many advancements in the study of biomaterials, device related infections remain a critical problem. To prevent bacterial biofilms from forming it is paramount to study and quantify the adhesion of bacteria onto various surfaces. Our lab leverages experimental thin film mechanics to investigate biological adhesion onto implants. We have adapted the laser spallation technique to determine dominant parameters that promote strong biofilm adhesion and are establishing a novel Adhesion Index – a ratio of mammalian cell adhesion to biofilm adhesion. With these new tools, we investigate the role of biofilm mechanics towards the propensity to develop medical device infections.

# 636194 Spatiotemporal Analyte and Metabolic Tracking in Microbial Biofilms with Nanosensors

**Abstract:**

For millennia humans have used analytical approaches to study what is happening in biological systems to better understand them. The typical approach is to pull out a sample and measure it using offline analysis. This works amazingly well for many applications - including clinical diagnostics, field sampling, and lab research. However, this approach can be more limited when it comes to understanding dynamics in smaller systems such as microbial biofilms. There are a few key challenges. 1) Taking a sample out from the system can perturb the sample itself. 2) For some analytes and samples you are interested in, you can't get a big enough sample to analyze. For example, if you are trying to sample between microbes there isn't much volume to sample. 3) It can be hard to get continuous data with sampling - so if you are interested in kinetic results or changes over time, sampling isn't optimal. 4) It is hard to get spatial imaging with sampling approaches - so spatial variations can get lost.

Our group is studying metabolic dynamics in microbial systems such as biofilms, and we accomplish this by flipping this sampling paradigm. Rather than bring the sample out of the system to the diagnostic, we bring the diagnostic to the sample. We developed luminescent nanoparticle-based sensors that are embedded throughout biofilms that we study. While this approach is applicable to other systems, I will talk about our efforts in monitoring *Pseudomonas aeruginosa*, where we use both typical lab strains and patient isolates from our clinical cystic fibrosis collaborators.

I will discuss our optical oxygen-sensitive nanosensors that were used to measure 3D oxygen gradients in *P. aeruginosa* biofilms with minimal disruption to the biofilm structure. We created and characterized optical nanosensors for detection of molecular oxygen which can be incorporated into the biofilm structure during growth. Using these, we obtained confocal microscopy data from which we can determine quantitative 3-dimensional oxygen gradients. We observed spatial gradation in oxygen concentration during biofilm growth, attributed to nutrient consumption at the edges of the biofilm. We also studied 3D oxygen gradients during antibiotic attack and found that oxygen was present at greater depths compared to untreated controls, consistent with cell death (as confirmed with follow-up plating). This approach can be used to determine strain specific metabolic responses to antibiotics – demonstrated with individual metabolic profiling of clinical isolates from three different patients. We can also use this nanosensor platform for higher throughput assays in multiwell plates to test antibiotic susceptibility in biofilms, measuring the strain dependent response to antibiotics in a simple assay. Overall, the use of optical nanosensors unlocks new areas of measurement in microbial systems to better study their function.

# 636260 Bacterial Biofilms Sense and Response to Substrate Stiffness

**Abstract:**

The ability of bacteria to colonize and grow on different surfaces is an essential process for biofilm development and depends on complex biomechanical interactions between the biofilm and the underlying substrate. Changes in the physical properties of the underlying substrate are known to alter biofilm expansion, but the mechanisms by which biofilms sense and respond to physical features of their environment are still poorly understood. Here, we report the use of synthetic polyacrylamide hydrogels with tunable stiffness and controllable pore size to assess physical effects of the substrate on biofilm development. Using time lapse microscopy to track the growth of expanding *Serratia marcescens* colonies, we find that biofilm colony growth can increase with increasing substrate stiffness on purely elastic substrates, unlike what is found on traditional agar substrates. Using traction force microscopy, we find that biofilms exert transient stresses correlated over length scales much larger than a single bacterium. Our results are consistent with a model of biofilm development in which the interplay between osmotic pressure arising from the biofilm and the poroelastic response of the underlying substrate controls biofilm growth and morphology.

# 625874 High-Throughput and Continuous Chaotic Bioprinting of Spatially-Controlled Bacterial Microcosms

**Abstract:**

Microorganisms do not work alone but instead function as collaborative micro-societies. The spatial distribution of different bacterial strains (micro-biogeography) in a shared volumetric space, and their degree of intimacy, greatly influences their societal behavior. Current microbiological techniques are commonly focused on the culture of well-mixed bacterial communities and fail to reproduce the micro-biogeography of polybacterial societies.

Here, we bioprinted fine-scale bacterial microcosms using chaotic flows induced by a printhead containing a static mixer. This straightforward approach (*i.e.*, continuous chaotic bacterial bioprinting) enables the fabrication of hydrogel constructs with intercalated layers of bacterial strains. These multi-layered constructs are used to analyze how the spatial distributions of bacteria affect their social behavior. For example, we show that bacteria within these biological microsystems engage in either cooperation or competition, depending on the degree of shared interface. The extent of inhibition in predator-prey scenarios (i.e., probiotic-pathogen bacteria) increases when bacteria are in greater intimacy. Furthermore, two *Escherichia coli* strains exhibit competitive behavior in well-mixed microenvironments, whereas stable coexistence prevails for longer times in spatially structured communities.

We anticipate that chaotic bioprinting will contribute to the development of a greater complexity of polybacterial microsystems, tissue-microbiota models, and biomanufactured materials.

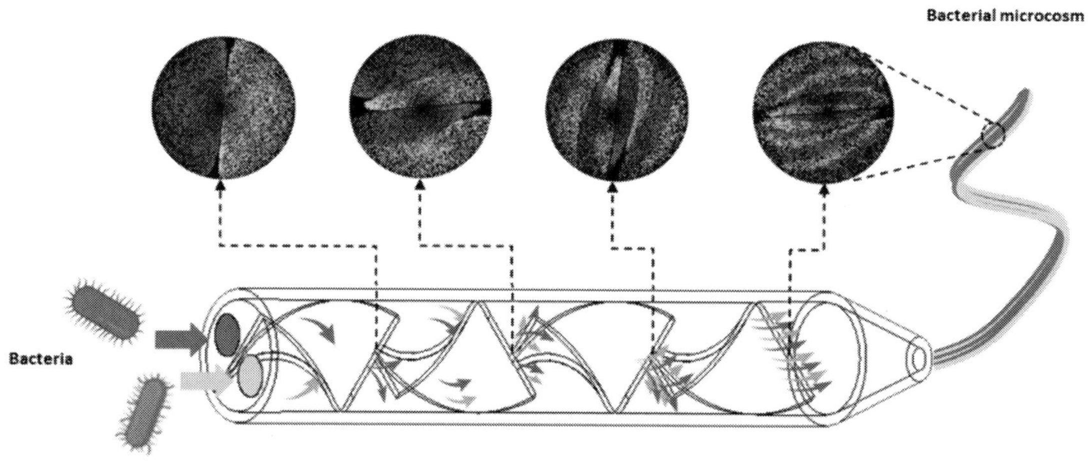

# 628641 Targeted Disruption of Dual Species Interfacial Films of *P. Aeruginosa* and *S. Aureus*

**Abstract:**

Chronic lung infection with bacterial biofilms is one of the leading causes of death in cystic fibrosis (CF) patients. Among many species colonizing the lung airways, *Pseudomonas aeruginosa* and *Staphylococcus aureus* are two of the most prevalent pathogens responsible for recalcitrant infections. These pathogens contribute to a mechanically robust biofilm that is difficult to eradicate using methods such as lung lavage and airway clearance techniques. Disrupting agents such as glycoside hydrolase-based compounds are commonly employed to target and break down the biofilm matrix components, and subsequently increase cell susceptibility to antibiotics. In this study, we utilized an interfacial system to study the effect of disrupting agents on the biomechanics of interfacial films. We hypothesize that our system will serve as a standardized platform to characterize and predict the effects of disrupting agents on the mechanical properties of interfacial films. We aim to characterize the effect of generic and targeted biofilm disrupting agents on the mechanical properties of single and dual species interfacial biofilms of *P. aeruginosa* and *S. aureus*. Specifically, we assessed viscoelasticity and mechanical properties of the interfacial films using pendant drop tensiometry and pendant capsule elastometry followed by the treatment with disrupting agents. Generic enzymes targeting the 1,4 linkages in the polymer backbone such as glycoside hydrolases (Cellulase and -amylase) and polymer specific enzymes such as DNase targeting the pel (pellicle) and eDNA complex in *P. aeruginosa* biofilms and dispersin B targeting the polysaccharide intercellular adhesion in *S. aureus* biofilms were compared for their efficiency to disrupt biofilms. Overall, precise prediction and characterization of the effect of biofilm disrupting agents will provide new insights towards the development of methods for treating interfacial biofilms.

# 631237 In Situ Mapping of the Mechanical Properties of Sulfate Reducing Bacteria By Microrheology

**Abstract:**

Sulphate-reducing bacteria (SRB) are corrosive anaerobic bacteria that secrete exopolymers on metal surfaces and form a biofilm to induce uniform corrosion or localized pitting. Quantifying the rheological properties of SRB biofilms at the micron scale is an important step to investigating the communal lifecycles of the bacteria as they adhere to substrate surfaces. The bacteria can modulate the mechanical properties of the biofilm according to the requirements of their colonial lifecycles. Rheology at the micron scale is directly relevant to the resistance of the biofilm from mechanical damage. Passive microrheology is a reasonably well-established technique that can be used to investigate the viscoelasticity of biofilms. Multiple particle-tracking (MPT) relies on tracking the Brownian motion of particles that are embedded in a medium to calculate the mechanical properties of the medium in the vicinity of the particles. By tracking the motion of the particles, measuring the viscoelasticity regarded by individual bacteria is possible. MPT resolves the local mechanical properties of the biofilm hydrogel spatially and temporally. We have developed a set of protocols that utilize MPT to characterize the mechanical properties of anaerobic bacteria in situ. To avoid biofilm disruption and eradication, fluorescent tracer particles are integrated into the media of the biofilms prior to inoculation. The primary contributors to biofilm mechanical properties, contributing to the overall viscoelastic character, have been investigated. The biofilm has been examined at different times after cultivation and under alternate environmental conditions. Our preliminary data confirms the establishment of biofilm growth, while showing different levels of heterogeneity in the biofilm.

# 632956 Polymer Surface Dissection for Characterization of Early Colonizing Microbes on Membrane Interfaces

**Abstract:**

Multispecies biofilms often develop over membrane surfaces in bioseparation systems, leading to membrane clogging and failure. This complex, multi-cellular assembly process is poorly understood from a mechanistic standpoint, in part due to the limited number of tools available for characterizing the cell-cell and cell-surface interactions that occur. Here, we report a high-resolution, materials-based approach for investigating spatiotemporal aspects of biofilm formation. The approach, termed Polymer Surface Dissection (PSD) relies on photodegradable polyethylene glycol (PEG) hydrogels to remove and isolate bacteria that are attached to membrane surfaces at different points in the fouling process for their molecular characterization. Hydrogels are functionalized with bio-affinity ligands to lift microbes off of membrane surfaces following contact. Subsequent exposure of detached cells within the hydrogels to patterned UV light enables their release into solution for retrieval, DNA extraction and 16S amplicon sequencing. The optimized PSD approach enables culture-free identification of membrane-bound colonies at areas as low as 2000 micron$^2$. With this approach, we investigate spatiotemporal aspects of biofouling from wastewater communities over PVDF membranes. Specifically, we identify sub-communities of microbes that are early colonizers on these membrane surfaces. As early-colonizing microbes initiate and accelerate the biofouling process, their identification will inform treatment strategies that can extend membrane lifetimes.

# 635450 Titanium Wires Modified By Peptoid-Loaded Microgels Resist *S. Aureus* Colonization

**Abstract:**

A wide range of surface-modification technologies continue to be explored to inhibit the bacterial colonization of synthetic surfaces with the aim of preventing infection associated with biomedical devices. We have been studying a self-defensive approach (1) where polyanionic microgels deposited electrostatically onto a synthetic surface are loaded with cationic antimicrobials which are later released upon exposure to bacterial triggering. We have previously demonstrated this process (2) with microgels based on poly(acrylic acid) [PAA] loaded with colistin or with an antimicrobial peptide. We are now exploring this concept using an antimicrobial peptoid known as TM1 ($M_W$=1819 Da, +5 charge). Peptoids are *N*-substituted glycines whose side chains are appended to the nitrogen rather than to the α-carbon of an amino acid (3). Here we study the complexation and sequestration properties of TM1 within PAA microgels synthesized by membrane emulsification coupled with free-radical (thermal or UV) polymerization. Complexation and sequestration can be followed using *in situ* optical microscopy to measure changes in the diameter of microgels immobilized on glass. When exposed to low-ionic-strength phosphate buffer containing dissolved TM1, the average microgel diameter rapidly decreases by ~60% due to osmotic effects associated with counterion release during TM1-PAA complexation and to the additional crosslinking caused by the polyvalent TM1 macro-ions. The average microgel diameter can remain unchanged for time scales of weeks when exposed to buffer (PBS) or culture medium, indicating that the TM1-PAA complexation is sufficiently strong to resist a burst release when exposed to physiological pH and ionic strength. We have challenged glass surfaces with *S. epidermidis* and used a PetriFilm assay to show that surfaces modified by TM1-loaded PAA microgels exhibit little or no subsequent bacterial growth while bacteria proliferate on unmodified control surfaces. We have furthermore developed an *in vitro* model of contamination within the operating room (OR) where bacteria can sediment from multiple sources - e.g. ventilation systems, shedding from clothing, staff sneezing/coughing, personnel traffic, etc. - onto exposed surfaces. As a substrate we used Ti wire typical of that used in an intermedullary rodent model and exposed wire samples to controlled bursts of aerosolized buffer containing known concentrations of viable methicillin-susceptible *S. aureus*. We studied Ti wires with four different treatments: unmodified, PAH-primed, (unloaded) microgel-modified, and TM1-loaded microgel-modified. Using assays to assess the presence and viability of both well-adhered and loosely bound bacteria, we find that TM1-loaded microgel-modified Ti wires effectively inhibit *S. aureus* colonization in this *in vitro* OR contamination model.

(1) X. Xiao, W. Zhao, J. Liang, K. Sauer, and M. Libera, *Self-Defensive Antimicrobial Biomaterial Surfaces.* Colloids and Surfaces B: Biointerfaces, 2020. DOI: 10.1016/j.colsurfb.2020.110989

(2) J. Liang, H. Wang, and M. Libera, *Biomaterial Surfaces Self-Defensive against Bacteria by Contact Transfer of Antimicrobials.*Biomaterials, 2019. **204**(June): p. 25-35.

(3) N. Chongsiriwatana, R.N. Zuckermann, A.E. Barron, et al. *Peptoids that Mimic the Structure, Function, and Mechanism of Helical Antimicrobial Peptides.* Proc Natl Acad Sci U S A, 2008. **105**(8): p. 2794-9.

# 635451 The Triggered Delivery of Polymyxin Antibiotics from Polyelectrolyte Microgels

**Abstract:**

One form of a self-defensive infection-resisting surface sequesters an antimicrobial within a coating until a bacterial challenge triggers release. The trigger is highly localized and can be caused by acid or enzyme secretion due to bacterial metabolism or simply by local thermodynamic changes induced by the highly anionic nature of a bacterial envelope. We are developing self-defensive surfaces using polyanionic microgels which can be electrostatically deposited onto almost any surface to form a sub-monolayer coating. Loading occurs by a subsequent self-assembly step where multivalent antimicrobial macro-cations complex within the polyanionic microgels. We have found that a broad variety of cationic molecules can be loaded from low-ionic-strength buffer where the electrostatic shielding effects of $Na^+$ are small. We have shown (1), for example, that colistin (polymyxin E; +5) can be electrostatically loaded from low-salt buffer into microgels made of poly(acrylic acid) (PAA), remain sequestered there for weeks, and kill *E. coli* on contact when exposed to a bacterial challenge. However, the colistin is released in an ineffective burst when the ionic strength of the surrounding buffer is increased to 0.14 M, because $Na^+$ shielding interferes with the colistin/PAA complexation. In contrast, the L5 (+6, 2274 Da) and Sub5 (+7, 1723 Da) antimicrobial peptides both remain sequestered in PAA under physiological conditions and are able to create surfaces that are self-defensive against staphylococci.

With the objective of creating self-defensive surfaces using FDA-approved antibiotics, we are working to identify the key properties that control microgel/antibiotic complexation and, thus, the ability of microgels to sequester and release their antibiotic payload. Here we focus on polymyxin B and E and their complexation with poly(styrene sulfonate) (PSS) microgels. We follow loading and sequestration using time-resolved optical microscopy to measure microgel deswelling/swelling in a continuous flow chamber that provides precise control over pH, ionic strength, and antibiotic concentration. Recent work (2) has shown that the PSS/colistin complexation strength is greater than the PAA/colistin complexation strength, an effect demonstrated by coarse-grain molecular dynamics simulations to be caused by the presence of the pendant phenyl rings along the PSS backbone. Except for a phenyl ring that substitutes for a dimethyl group, polymyxin B is chemically identical to colistin, yet we find that the threshold $[Na^+]$ in phosphate buffer to trigger release (de-complexation) is 0.216 M for colistin and 0.516 M for polymyxin B release, again indicating that aromaticity plays an important role in complexation and antibiotic sequestration. This finding has important implications on how each of these antibiotics might be incorporated into a self-defensive technology, and work continues to assess the interactions of polymyxin-loaded PSS microgels with gram-negative bacteria.

(1) J. Liang, H. Wang, and M. Libera, *Biomaterial Surfaces Self-Defensive against Bacteria by Contact Transfer of Antimicrobials.*Biomaterials, 2019. **204**(June): p. 25-35.

(2) X. Xiao, J. Ji, W. Zhao, S. Nangia, and M. Libera. *Salt Destabilization of the Complexation of Cationic Colistin within Polyanionic Microgels*, submitted to Macromolecules, 2021.

# 628902 Advancing the Functionality of Dimethylallylamine-Based Nanocultures for Selective Transport and Controlled Microbial Dynamics.

**Abstract:**

The need for new assessment tools for microbial dynamics has necessitated the miniaturisation of cell-culturing techniques and the design of microsystems that facilitate the interrogation of microorganisms in-well-defined environments. The nanocultures, as described in this work, are a type of such assessment tool: nanolitre-sized microcapsules generated using a flow-focusing microfluidic device to sequester and cultivate microbes in a high-throughput manner. By manipulating the chemistry of their polymeric shell, the nanocultures can be designed to achieve new functionalities, such as a selective permeability facilitating the transport of metabolites, nutrients and other small molecules essential to control cell growth and characterize community dynamics. In this work, transport properties of a newly designed Poly(dimethylsiloxane)-based polymer membrane functionalized with Dimethylallylamine (DMAA) have been examined by investigating the diffusion of selected molecules relevant to controlling cell dynamics, including antimicrobials, fluorescent staining probes and sugars. Furthermore, the Flory-Huggins Interaction Parameter was evaluated as a predictive tool to elucidate the partitioning and transport of selected molecules into the nanocultures. Diffusion of molecules was confirmed experimentally by generating nanocultures containing *Escherichia coli* cells, whereby cell growth was used as a proxy for determination of successful molecule diffusion. These data will pave the way in our understanding of effectively using the nanocultures to study complex synergistic and antagonistic microbial behaviours in both natural and synthetic communities, with the goal of better simulating natural microenvironments and increasing discoverability of novel molecules such as antimicrobials.

# 622959 Biofilm Reduction Via Surface Vs. Immersion Heating

**Abstract:**

Despite tremendous breakthroughs in the development of medical implants, biofilm infections on these devices remain a perennial problem. One approach to biofilm destruction is localized thermal shock, which triggers both dispersion of bacteria from the biofilm and death of bacteria remaining in the biofilm. The relative magnitudes of these phenomena may be difficult to compare, however. Most reports on biofilm thermal susceptibility are from immersion in thermostatted media as this enables prompt and uniform delivery of thermal shocks for biofilm destruction. However, this approach also kills much of the dispersed bacteria, which cannot then be readily quantified, making the magnitude of dispersed bacteria unknown. Although dead bacteria in the biofilm can be quantified arithmetically via confocal microscopy, this approach is inaccurate and cannot be used for quantification of dead bacteria in the media. Moreover, delivering thermal shocks via immersion is only possible in limited circumstances clinically (e.g., within the lumen of a catheter). Therefore, this study investigates the efficacy of surface heating for biofilm reduction.

Mature *Pseudomonas aeruginosa* biofilms cultured on thermoelectric devices were thermally shocked at 37, 60, 70, or 80 °C for 1, 5, or 30 min in the presence of 5 ml 3 g/l tryptic soy broth. Quantifying bacterial population of the media immediately after the thermal shocks showed that there is a roughly constant ratio between the population density on the surface and in the media throughout all the trials, including controls with no thermal shock. This suggests that dispersion is an equilibrium process and thermal shock does not alter this equilibrium significantly. This shows that bacterial reduction by thermal shocks can be achieved by bacterial death in either the biofilm or the surrounding media. Furthermore, the efficacy of surface heating in bacterial reduction was higher at shortest (1 min) thermal shocks compared to immersion heating. However, surface heating was less effective than immersion heating when shock exposure increased to 5 or 30 min. The decrease in efficacy at longer times may be due to bacteria partitioning into the cooler media, where they experience less thermal damage, before returning to the biofilm to maintain the population equilibrium between the phases. This hypothesis was tested by replacing the liquid media above the biofilm with hydrogel tissue mimic to inhibit dispersion from the biofilm.

# 623373 A New Antimicrobial Peptide-Antibiotic Combination Strategy for *P. Aeruginosa* Inactivation

**Abstract:**

*Pseudomonas aeruginosa* (*P. aeruginosa*) is an opportunistic pathogen which causes several types of severe infections in humans. Traditional antibiotic treatments are not always effective at curing these infections because the lower outer membrane permeability of *P. aeruginosa* limits the uptake of antibiotics. In addition, many bacteria are now resistant to the commonly used antibiotics for *P. aeruginosa* infections. Currently, antibiotic combinations are administered to treat *P. aeruginosa* infections, such as Ceftolozane-Tazobactam or Ceftazidime-Avibactam in combination. However, there are concerns that the antibiotic combination may accelerate the already existing drug resistance. We propose developing a novel treatment strategy through combining the human antimicrobial peptide LL-37 with different antibiotics, to combat *P. aeruginosa* strains *in vitro*. Preliminary results showed that LL-37 exhibits synergistic effects against *P. aeruginosa* when combined with some traditional antibiotics. Further, the effective concentrations of LL-37 and antibiotics decrease substantially compared with the original minimum inhibitive concentrations, suggesting that there is the potential to decrease the cytotoxicity during treatment. In addition, some antibiotics we used in our research are not considered as options to treat infections in clinical due to their weak penetration abilities to the outer membrane of *P. aeruginosa*. It is therefore promising that they were effective in the combination strategy. We infer that LL-37 helped to increase the outer membrane permeability of *P. aeruginosa* and thus enhance the uptake of antibiotics, which makes these antibiotics effective against *P. aeruginosa*. Further research will be conducted to help elucidate the mechanism of these synergistic effects.

# 625521 A Targeted Genome-Wide Approach to Elucidate and Control Bacterial Adhesion to Physicochemically Diverse Biomaterials

**Abstract:**

Biofilm infections on indwelling biomedical devices are the leading cause of hospital infections around the world. The most common are catheter-associated infections caused by *Escherichia coli* and *Staphylococcus aureus*. Prevention and treatment of these infections typically rely on broad-spectrum antibiotics, yet these strategies often promote antibiotic resistance and are not fully effective. Elucidating the genetic elements responsible for bacterial adhesion to and biofilm formation on biomaterials could be used for the rational design of targeted antibiofouling agents that prevent infections on biomedical device surfaces. In this work, we aim to systematically map the genomes of *E. coli* and *S. aureus* to identify genetic targets that confer adhesion phenotypes and quantify their fitness using a multiplexed, high-throughput approach. In this approach, we design comprehensive genome-wide CRISPR interference (CRISPRi) gene repression libraries for *E. coli* and *S. aureus* and apply pooled selections on a platform of biomaterials (PEG and PDMS gels) that systematically vary in their physicochemical properties to uncover gene expression associated with bacterial attachment to various surfaces. We have developed and characterized CRISPRi tools in *E. coli* MG1655 using three CRISPRi systems to determine their gRNA design rules and appropriate expression levels of each component to prevent cellular toxicity. Repression of genomic loci, including loci associated with biofilm formation, was also tested and quantified using superfolder GFP fusions. Using the determined design rules and custom Python scripts, we have designed gRNA libraries (>40,000 designs each) for each CRISPRi system to target nearly every annotated gene in the MG1655 genome (95.9%–99.1%). For *S. aureus*, very few synthetic DNA parts have been characterized and reported in the literature. Here, we created and report what we believe to be the first characterized genetic part toolbox for *S. aureus*, including promoters, ribosome binding sites, terminators, and plasmid origins of replication. We then used the genetic toolbox to rationally design CRISPRi tools for *S. aureus*. These genome-wide CRISPRi tools can be applied to study genotype-phenotype relationships under different conditions and can be developed for any genetically tractable organism with an annotated genome.

# 630624 Establishing Relationships between Local pH and Mechanics in *Staphylococcus Epidermidis* Biofilms

**Abstract:**

Bacterial biofilms are complex, structured communities of microbial cells encased within self-produced extracellular polymeric matrix materials. *Staphylococcus epidermidis* biofilms frequently cause infections on catheters and implanted medical devices, and are responsible for nearly 120,000 infections annually[1]. As *S. epidermidis* biofilms grow, acidic waste products such as lactic acid, acetic acid, and formic acid can accumulate within their structures. Additionally, *S. epidermidis* biofilm matrix materials have been shown to be unstable and more rigid at lower pH values and stable and softer at higher pH values[2]. We hypothesized that local pH gradients and heterogenous local mechanics exist within *S. epidermidis* biofilms due to the structural heterogeneity and nonuniform distribution of acidic waste products in biofilms. To test this hypothesis, we used confocal laser scanning microscopy (CLSM) and image analysis techniques to quantitatively map local pH gradients and determine local biofilm mechanical properties. The most acidic microenvironments were found to be at the biofilm-substrate interface during biofilm growth in both static culture and at physiological shear stresses. Local mechanical properties were more heterogeneous in biofilms grown at physiological shear stresses than in static culture. *S. epidermidis* biofilm mechanics varied locally with pH in both static growth environments and at physiological shear stresses, where local pH $> 6.9$ within biofilms resulted in softer mechanics than in biofilm regions with lower average pH. This work advances how variations in the local microenvironment can shift biofilm mechanics and has implications for the development of biofilm disruption techniques targeting local biofilm mechanical properties.

[1]Kourtis, A. P. et. al. (2019). *Morbidity and mortality weekly report*, *68*(9), 214–219.

[2]Stewart, E. J., Ganesan, M., Younger, J. G., & Solomon, M. J. (2015). *Scientific reports*, *5*, 13081.

# 624559 Design of a Well-Defined Poly(dimethylsiloxane)-Based Microbial Nanoculture System

**Abstract:**

Organosilanes contain hydrocarbon-like backbones, allowing them to react with silicone-based agents in the presence of a catalyst and polymerize into membranes with tunable transport and mechanical properties. Owing to their high hydrophobicity, Poly(dimethylsiloxane) (PDMS) membranes, and more particularly, Sylgard® 184, have been used for applications including drug delivery, gas separation, and microfluidics fabrication. However, the undefined composition of the material and its ability to leach out uncured oligomers make its functionalization and usage challenging for many biological applications. This article presents the design of a novel culture system generated using PDMS-based membranes to study microbial dynamics. The microbial culture system that is referred to as "nanoculture" serves to encapsulate and grow microbes in semipermeable membranes. The mechanical properties of the membranes are investigated through osmotic annealing to determine the ability of the nanocultures to withstand high shear stress similar to environmental conditions while maintaining transport properties essential to microbial communication and growth. The present study lays the foundation for a novel microbial culture system that would enable the cultivation of microorganisms in environments other than laboratory conditions.

# 622661 To Biofilm or Not to Biofilm: Interplay between Chemotactic Dispersal and Biofilm Formation in Bacterial Communities

**Abstract:**

Bacteria are ubiquitous in our daily life, existing in soils, sediments, and even our body, frequently as surface-attached biofilm communities. In some cases, biofilms serve a positive purpose, such as improving health or remediating polluted water; in other cases, they negatively impact our lives, such as by causing infection or fouling equipment. For both positive and negative purposes, understanding the factors that regulate the onset of biofilm formation is crucial in determining how to control or treat them. However, how bacteria transition between the free-swimming planktonic state to the sedentary biofilm state in these heterogeneous environments is poorly understood. Here, we use computational modeling to investigate how biofilm formation depends on bacterial properties as well as the properties of their environment. Specifically, by analyzing the competition between chemotactic dispersal and quorum sensing, we establish universal rules predicting how the onset and extent of biofilm formation depend on cell concentration and motility; nutrient, diffusion, and consumption; chemotactic sensing; and autoinducer secretion. The findings from this study therefore yield quantitative principles by which biofilm formation can be predicted and controlled.

# 634816 Invited Talk: Taking Electrochemical Biomolecular Sensors from the Bench to the Clinic

**Abstract:**

To put disease-related biomarkers to work in the clinic, new high-performance technologies are needed to enable rapid and sensitive analysis of clinical specimens. Electrochemical methods providing low cost and direct biomarker readout have attracted a great deal of attention for this application. In our laboratories, we have used controlled nanostructuring of electrode surfaces to enhance biomolecular capture rates and efficiencies (*J. Am. Chem. Soc.* 2017). By immobilizing biomolecules on nanostructured surfaces, high-performance biomolecular detection systems can be developed to facilitate rapid biomarker analysis and the detection of infectious pathogens (*J. Phys. Chem. C.* 2016; *Nano Lett.* 2017). We have developed electrochemical assays that are able to detect nucleic acids, proteins and small molecules, with universally high sensitivity levels and have applied this approach in a variety of areas including noninvasive monitoring via Liquid Biopsy (*Nature Chemistry* 2015) and antimicrobial resistance assessment (*Nature Chemistry* 2020). Recently we have developed reagentless sensors that are powerful detectors for viral sensing (*J. Am. Chem. Soc.* 2021) and in vivo sensing (*Nature Chemistry*, 2021). This talk will summarize the development of these sensors, their application to a variety of clinical problems, and their evolution into a clinical-grade products.

# 627855 Sensitive Electrochemical Detection of Pathogenic *E. coli*

**Abstract:**

Pathogenic *E. coli* are a significant threat as a foodborne pathogen and a cause of urinary tract infections. Conventional methods to detect these microbes require centralized facilities and trained personnel, as well as hours to days to complete. We have developed a sensitive electrochemical platform to covalently capture, detect, and quantify viable pathogenic *E. coli* from complex matrices with a detection limit of 12 CFU/mL and a linear range of detection up to $10^7$ CFU/mL. The technology enables quantification within two hours on disposable electrodes, even from milk and artificial urine. Endogenous *E. coli* were successfully detected from commonly-contaminated samples, including eggs, raw chicken, spinach, romaine lettuce, and an infected urine sample. The *E. coli* quantification was as sensitive as the current gold-standard of colony counting, but with significantly less error. This platform represents a major step in the development of field-deployable sensors to detect these dangerous microbes.

# 628278 Synthetic Biology Mediated Electrochemical Sensing Strategy

**Abstract:**

Synthetic biology enables integration of biomolecules with different functions in a modular manner toward desired applications. However, the output signal of current biochemical circuits primarily remains optical. We herein present strategies to apply electrochemistry to evaluate the performance of multi-function biochemical circuits. A CRISPR mediated primer-exchange-reaction based genetic circuit is designed toward probing specific regions of the genome of SARS-CoV-2. The output of the gene circuit is connected with a single-use, microfabricated gold sensing surface, enabling the direct translation of biomolecular information into electrical signal. The integration of electrochemistry into synthetic biology can lead to a new paradigm for sensor development and can also be a high-throughput interface for the investigation of large-scale gene circuits.

# 631801 Hydrogel-Encapsulated Gold Nanoshells Prepared By Inverse Emulsion Polymerization As a Biosensor for Sjögren's Syndrome Protein Markers

**Abstract:**

Proteomics research has led to identification of increasing numbers of differences in the protein makeup of biological fluids, such as saliva, tears, urine, or serum, due to the onset of disease; however, more work is needed to improve the molecular recognition capabilities of existing biosensors before these markers can see more widespread adoption in clinical practice. One approach for more robust, low-cost recognition elements is synthetic receptors based on polymeric hydrogels, but this approach is challenged by the lack of specificity toward the target analytes, which reduces their viability for recognition in biological fluids due to a myriad of competing molecules. One solution is differential sensor arrays that exploit semi-selective recognitive elements in order to screen for differences in multiple protein biomarkers in a single assay. In this work, we explore the capabilities of using inverse emulsion polymerization to form libraries of multifunctional hydrogel networks on the surface of gold nanoshells (AuNS) as semi-selective recognition elements for use in a localized surface plasmon resonance-based biosensor for the detection of biomarkers associated with Sjögren's syndrome.

Hydrogel nanoparticles were prepared by an inverse emulsion copolymerization of methacrylic acid with acrylamide or *N*-isopropylacrylamide, crosslinked with *N,N'*-methylene bisacrylamide. The particles were characterized by dynamic light scattering, potentiometric titration, and Fourier-transform infrared spectroscopy. AuNS were prepared using the method of seeded growth of gold colloids on aminated silica nanoparticles followed by encapsulation with poly(maleic anhydride-*alt*-1-octadecene)-g-poly(ethylene glycol) methacrylate (PMAO-g-PEGMA) graft copolymer to improve particle stability. Hydrogel-coated AuNS were prepared with the same procedure as the nanoparticle synthesis with the addition of PMAO-g-PEGMA AuNS to the aqueous phase of the inverse emulsion. UV-Vis spectroscopy was used to measure the shifts in the wavelength of the localized surface plasmon resonance for the hydrogel-coated AuNS. Inverse emulsion synthesis was found to lead to thin surface coatings for both formulations and shows immense promise for the creation of a library of hydrogels with various functional groups that could be used to alter affinity for different protein biomarkers from patient biological fluids.

Work supported by National Institutes of Health Grant R01-EB022025

# 628700 Silk-Based Microneedle Biosensor for Sustainable Food Supply Chain

**Abstract:**

Manipulation of hierarchical assembly of biopolymers enables utilization of their functionalities for engineering purposes. Silk fibroin extracted *Bombyx mori* cocoons is an emerging technical material for food contact applications due to its non-toxic and edible nature. In this study, we present a silk-based microneedle injection system for food bacteria and spoilage detection. We developed internal pore structures of microneedles through the water-annealing process leading to increased beta-sheet formation. The microneedles delivered internal food fluid via capillary action in pores to a colorimetric sensor on the needle array's backside. Within a day of injection, we identified *E. coli* contamination and common food spoilage in fish fillets by color changes of the polydiacetylene-based sensor. We also showed that the microneedle sensor could pierce commercial food packaging, implying a successful adaptation of the technology downstream in conventional food supply chains. Therefore, this study provides insights into global food safety and food waste reduction.

# Engineered CRISPR-ENHANCE system for clinical detection of SARS-CoV-2 RNA

Long T. Nguyen, Santosh R. Rananaware, Brianna L.M. Pizzano, Brandon T. Stone, Piyush K. Jain*

*Corresponding and presenting author

The Severe Acute Respiratory Syndrome-Coronavirus-2 (SARS-CoV-2) has resulted into 2019 novel coronavirus disease (COVID-19) pandemic with confirmed almost 3 million deaths and over 136 million confirmed cases worldwide as of April 12th 2021. There is an urgent need for the development of a rapid, sensitive, specific, inexpensive, easy-to-use, and accessible testing kit for SARS-CoV-2 that can be implemented as a point-of-care diagnostic and a home-based testing kit. Most Class 2 CRISPR/Cas systems, including CRISPR/Cas12 (Type V), CRISPR/Cas14 (Type V), and CRISPR/Cas 13 (Type VI) mediate a nonspecific collateral trans-cleavage of random DNA and RNA after binding or cis-cleavage of their target DNA or RNA. For CRISPR/Cas12a, this multiple-turnover trans-cleavage activity is only initiated once the crRNA/Cas12a complex is bound to its target ssDNA or dsDNA that acts as an activator. This trans-cleavage activity has been widely exploited for nucleic acid detection and has been combined with fluorescence-based, paper-based, and electrochemical-based sensing technologies to develop rapid and sensitive diagnostics. Therefore, Type V and VI CRISPR/Cas systems have recently emerged as diagnostics for detecting nucleic acids including SARS-CoV-2 RNA.[1-11] However, current CRISPR/Cas12a systems are limited to a picomolar detection limit unless the target is either pre-amplified in some manner. The Cas12a-based DETECTR technology from Mammoth Biosciences and Cas13a-based SHERLOCK technology from Sherlock Biosciences have been approved by the FDA under EUA as lab-based diagnostics for detecting SARS-CoV-2 genomic RNA in a fluorescence-based assay.

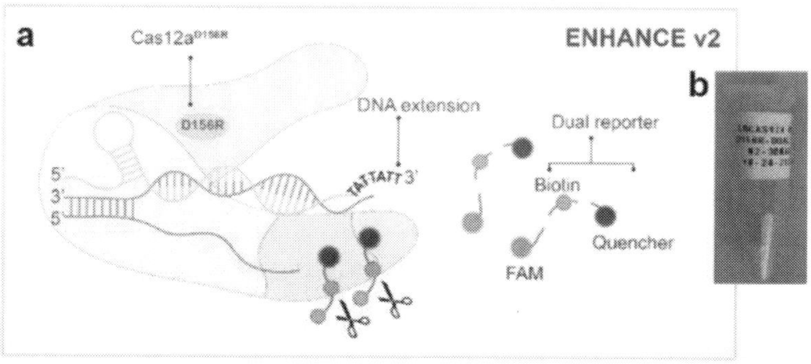


**Figure 1**: *Development and validation of a lyophilized and lateral flow version ENHANCEv2. (a) ENHANCEv2 is comprised of an engineered LbCas12a with D156R mutation, crRNAs containing 7-nt DNA extensions on 3' end, and a dual reporter. Once a dsDNA target is recognized by ENHANCEv2, it cleaves dual reporters and allows for enhanced multimodal detection using a lateral flow assay and a fluorescence-based assay in a single assay. (b) A lyophilized version of ENHANCEv2 demonstrated enhanced stability at room temperature for at least 30 days.*

We recently discovered that crRNAs can tolerate various length extensions with DNA or RNA on both 3'- and 5'-ends, while the phosphorothioate extensions reduced or eliminated the trans-cleavage activity (Nguyen et al., *Nature Communications*, 2020)[9]. In particular, an engineered crRNA containing a 7-nt DNA extensions on its 3'-end, drastically increased the trans-cleavage activity of CRISPR/Cas12a and increased specificity of detection. By extending the 3'- or 5'-ends of the crRNA with different lengths of DNA, RNA, and phosphorothioate DNA, we discovered a new self-catalytic behavior and an augmented rate of Cas12a-mediated trans-cleavage activity as high as 3.5-fold compared to the wild type crRNA, making it the fastest reported CRISPR/Cas12a in terms of trans-cleavage activity, and termed it as CRISPR-ENHANCE. This in turn reflected an unprecedented improvement in sensitivity and limit of detection of nucleic acid targets down to the femtomolar level for a variety of clinically-relevant nucleic acid targets including prostate cancer antigen 3 (PCA3), HIV, HCV, and SARS-CoV-2 RNA without target pre-amplification within 30 minutes. By combining with a reverse transcription-loop-mediated isothermal amplification (RT-LAMP) step, 100 aM of SARS-CoV-2 RNA could be detected using a lateral flow assay within 60 minutes (Nguyen et al., *Methods*, 2021)[10].

By further optimization of crRNAs and combining with a target pre-amplification strategy using a reverse transcription-loop-mediated isothermal amplification (RT-LAMP) step, a single molecule detection of SARS-CoV-2 RNA in nasopharyngeal swabs was achieved within 50 minutes. Compared to the wild-type crRNAs, our CRISPR-ENHANCE system demonstrated up to 5-fold higher sensitivity in detecting SARS-CoV-2 RNA. By combining engineered crRNAs with a Cas12a mutant (D156R), a multimodal reporter, and lyophilizing the components, the test could be conducted within 33 minutes using both a fluorescence-based assay and a lateral flow assay. The CRISPR-ENHANCE demonstrated high inclusivity and exclusivity and achieved 96.7% sensitivity and 96.7% specificity in clinical samples and was stable for at least 30 days at room temperature (Figure 1, Nguyen et al., *MedRxiv*, 2021- *in revision at Communications Medicine-Nature*)[11].

## References


1. Broughton JP, Deng X, Yu G, et al. CRISPR-Cas12-based detection of SARS-CoV-2. *Nat Biotechnol.* Apr 2020;doi:10.1038/s41587-020-0513-4
2. Abudayyeh OO, Gootenberg JS, Kellner MJ, Zhang F. Nucleic Acid Detection of Plant Genes Using CRISPR-Cas13. *CRISPR J.* Jun 2019;2:165-171. doi:10.1089/crispr.2019.0011
3. Kellner MJ, Koob JG, Gootenberg JS, Abudayyeh OO, Zhang F. SHERLOCK: nucleic acid detection with CRISPR nucleases. *Nat Protoc.* Oct 2019;14(10):2986-3012. doi:10.1038/s41596-019-0210-2
4. Gootenberg JS, Abudayyeh OO, Kellner MJ, Joung J, Collins JJ, Zhang F. Multiplexed and portable nucleic acid detection platform with Cas13, Cas12a, and Csm6. *Science.* 04 2018;360(6387):439-444. doi:10.1126/science.aaq0179
5. Myhrvold C, Freije CA, Gootenberg JS, et al. Field-deployable viral diagnostics using CRISPR-Cas13. *Science.* 04 2018;360(6387):444-448. doi:10.1126/science.aas8836
6. Gootenberg JS, Abudayyeh OO, Lee JW, et al. Nucleic acid detection with CRISPR-Cas13a/C2c2. *Science.* 04 2017;356(6336):438-442. doi:10.1126/science.aam9321
7. Aquino-Jarquin G. CRISPR-Cas14 is now part of the artillery for gene editing and molecular diagnostic. *Nanomedicine.* 06 2019;18:428-431. doi:10.1016/j.nano.2019.03.006
8. Chen JS, Ma E, Harrington LB, et al. CRISPR-Cas12a target binding unleashes indiscriminate single-stranded DNase activity. *Science.* 04 2018;360(6387):436-439. doi:10.1126/science.aar6245
9. Nguyen LT, Smith BM, Jain PK. Enhancement of trans-cleavage activity of Cas12a with engineered crRNA enables amplified nucleic acid detection. *bioRxiv.* 2020:2020.04.13.036079. doi:10.1101/2020.04.13.036079
10. Nguyen LT, Gurijala J, Rananaware SR, Pizzano BLM, Stone BT, Jain PK. CRISPR-ENHANCE: An enhanced nucleic acid detection platform using Cas12a. *Methods.* Feb 2021;doi:10.1016/j.ymeth.2021.02.001
11. Nguyen LT, Rananaware SR, Pizzano BLM, Stone BT, Jain PK. Engineered CRISPR/Cas12a Enables Rapid SARS-CoV-2 Detection. *MedRxiv.* 2020;doi:10.1101/2020.12.23.20248725

# 633118 Computational Design of MOF-Based Electronic Noses for Disease Detection By Breath

**Abstract:**

Although it well-known that our breath contains a multitude of biomarkers for various diseases, breath-based screenings and diagnostics are almost non-existent in a clinical setting. This is largely due to the fact the current state-of the-art gas sensing technologies, namely gas chromatography-mass spectrometry (GCMS), are too bulky and too expensive for screening a large number of samples. Furthermore, due to the sheer number of compounds in breath, simple low-cost sensors such as mass-based and chemiresistive devices struggle to differentiate between the compound(s) of interest and other unknown gases. Thus, conventional diagnostics such as blood tests are still preferred. However, researchers have envisioned using gas sensing arrays, often called electronic noses, to achieve a level of sensitivity and selectivity comparable to GCMS, while maintaining the cost, speed, and size advantages of simple sensors, ultimately creating a device which would make breath-based screening and diagnostics competitive with conventional techniques.

Herein, we discuss the use of metal-organic frameworks (MOFs) in electronic noses of surface acoustic wave (SAW) devices for breath-based disease detection. Owing to the number of MOFs (>130,000) one could choose from, as well as the number of gases found in our breath (700+), the task of building an electronic nose is as much of a big data problem as it is an engineering challenge. Thus, to properly address this big data aspect of the problem, our efforts to design an electronic nose have been computational in nature. Using Grand Canonical Monte Carlo (GCMC) simulations, we can generate massive libraries of adsorption data which can be used to match against real sensor outputs, enabling quantification of a gas mixture. However, as we desire to include more gases and detect them in smaller concentrations, the number of gas mixtures which would need to be simulated grows exponentially. Thus, in order to maintain a reasonable computational demand, we pivoted to a new model in which we use GCMC simulations to generate a library of adsorption coefficients which enable us to determine the adsorbed mass of trace gas species without running a distinct simulation for each new set of concentrations. In turn, we created a flexible algorithm for the quantification of trace gas species. Since most of the gases of interest in breath are present in trace quantities, the above method works very well for simultaneously quantifying numerous compounds in breath, and using that information for disease screening and diagnostics.

# 629784 Interaction of Acetylcholinesterase with Phorate

**Abstract:**

Detection of residual organophosphates in food is of paramount importance in current scenario. Biosensors for pesticide detection may be developed on solid surfaces by monitoring the interaction of acetylcholinesterase with pesticides. Acetylcholinesterase is a hydrolytic enzyme which hydrolyzes acetylcholine, a neurotransmitter, into choline and acetate. This enzyme gets inhibited in the presence of pesticides. Utilizing this property, protein based sensor on solid surfaces may be developed. The interaction studies were carried out on silica surface using acetylcholinesterase and phorate, an organophosphate. Self-assembled monolayer was created by immobilization of enzyme on silica surface and further pesticide is allowed to interact with adsorbed protein. Binding of immobilized protein and phorate on surface was evaluated in terms of the mass and thickness change of surface using quartz crystal microbalance which is calculated from frequency and dissipation factor change using Sauerbrey equation. Low value of dissipation factor ($\Delta D$) corresponds to stronger binding of pesticide with protein. The nature of chemical reaction between acetylcholinesterase and phorate was analyzed by the stoichiometry and energy released. The interaction study was carried out at low phorate concentration of 5 ppm of phortae. These findings helps in the development of protein based pesticide detection sensor so as to provide user friendly technique for detection in raw food samples.

**Keywords:** Acetylcholinesterase, ITC, pesticide, QCM.

**References:**

1. Singh, A. K. *et al.*, 1999 Development of sensors for direct detection of organophosphates. Part I: immobilization, characterization and stabilization of acetylcholinesterase and organophosphate hydrolase on silica supports. Biosensors & Bioelectronics. 14:703–713.

# 627267 Hyperspectral Imaging Sensors Enhance on-Line Food Quality and Authenticity Inspection.

**Abstract:**

Used extensively as a Remote Sensing technique in Agriculture and Environmental monitoring, Hyperspectral Imaging (HSI) makes rapid advances in industrial process monitoring and shows great potential to become a valuable technique for rapid qualitative and quantitative assessment of food quality, safety, and authenticity.

Hyperspectral Imaging is a technique that collects spectral information, resolved over 100+ bands, for every pixel in an image. This wealth of spectral and spatial information acquired during a scan can be used to solve challenges like detection of foreign objects or discrimination between types of objects for various applications.

Modern HSI cameras and integrated systems, coupled together with image processing and statistical tools, can be used for product or process characterization and for delivering quantified results.

Significant value can be realized by using HSI as a tool for early inspection at receiving stations in food processing facilities. Here the ability to quickly scan large quantities of product and get results without the need to prepare samples for laboratory analysis can help the user reduce the cost of quality. Better informed decisions can be made early in the process, helping the user assign value to a specific batch or decide that a certain batch does not meet quality criteria. The information from an early inspection can also be used to help improve quality upstream – in the fields, groves and farms.

An example of this type of HSI application is prediction of sweetness or Brix content in oranges. Brix value is one of the most important quality parameters for growers and processors and traditionally requires a sample to be sent to a lab, oranges to be juiced and soluble solids content measured using a refractometer. A case study described in this presentation shows a prediction of Brix based on HSI scan of a few oranges. Refractometer data was used as a reference. Results of the prediction suggest that the technique could be used for quality assessment of samples under controlled conditions. Methodology used to develop a spectral classification model also helped in creating an optimized model for run-time spectral classification. This enabled the system to produce real time Brix results for batches of oranges passing under the scanner.

Another area where HSI delivers significant value is food fraud detection. The technique is well established in forensic applications and art authentication but now, with reduced costs, simplified user interfaces and AI powered algorithms, HSI becomes a usable tool for industrial QC. Food fraud is a major global problem. In particular, products that move through global supply chains and are sold in powder or small flake form are susceptible to foul play. Oregano, for example can be adulterated with cheaper ingredients like myrtle or olive leaf. Detecting this visually is quite difficult and processing facilities often resort to tedious, manual inspection under a microscope or sending samples for a chemical analysis. With HSI the user can quickly scan a sample of the product and develop a model that would differentiate a pure sample from an adulterated one. Further image processing can help in producing quantified results as in % purity, but even a simple color classification can help make a quick decision if the sample is good or if it requires a more detailed analysis.

An important enabler for Hyperspectral technology in industrial applications is the software for real time spectral classification. Machine Learning algorithms allow even a non-specialist user to create practical, deployable models in a matter of hours. Important progress is also being made in sensor design, taking advantage of increasing power of embedded image processing computers complete hyperspectral systems can now be packaged into compact, rugged housings making the technology useful as a process analytical tool. Embedded computing and on-board processing or raw hyperspectral data overcomes one if the major historical hurdles for adoption of HSI – enormous quantities of raw data generated by the cameras. Modern integrated HSI systems with embedded processing can perform run-time spectral classification and produce an output that consists of a waterfall image stream, object metadata or predicted quality attributes like moisture content, Brix and other results of regression modeling.

processing technology in the Hyperspec MV.X is removing the complexity associated with handling vast amounts of raw data. By performing spectral classification and post processing images on board the device we can output actionable results – object locations, size or predicted values directly to the end user.

This solution produces a complete, fully integrated system in a single ingress protected housing. The design lends itself well to being integrated with other automation or operated as a standalone quality monitoring system. Networking capabilities of embedded computing turn it into a smart device that can be deployed as part of a digital solutions strategy. The scalability of having the same system suitable for both small scale bench testing and for full on-line production environment is a big value add for the users.

# 621789 Collagen Type II Quantification with Pd@Pt Nanoparticle-Linked Immunosorbent Assay

**Abstract:**

In this presentation, we show a method for quantitatively detecting a larger-sized analyte, Collagen type II, using a Pd@Pt nanoparticle-linked immunosorbent assay.

Extracellular matrix (ECM), including cartilage tissue, is made up of proteoglycan and collagens, which act as the main structural proteins. Collagen types I, II, and III are essential fibril-forming collagens that comprise 80 – 90% of all collagens found in the human body. Collagen type II makes up approximately 50% of the protein and 85 – 90% of collagen in articular cartilage. Articular cartilage is the tissue that connects diarthrodial joints, providing a cushion between the bones.

The Centers for Disease Control and Prevention (CDC) estimates that over 32.5 million US adults suffer from arthritis. The most common type of arthritis, osteoarthritis, is a degenerative joint disease frequently found in the hands, hips, and knees. Osteoarthritis causes symptoms such as pain, swelling, and less flexibility because of the erosion and damage of articular cartilage. When osteoarthritis is present in a joint, components of the ECM, Collage type II and aggrecan, are broken down by the enzyme collagenase. Therefore, Collage type II quantification is a potential method for observing the degree of cartilage damage in osteoarthritis patients.

There presently is no treatment or cure for osteoarthritis, which is why we are studying articular cartilage regeneration through the use of scaffold-based techniques. Collagen type II is one component of the regenerative medical scaffolding used for cartilage and tissue engineering applications. Therefore, the ability to produce Collage type II and quantify the production is needed.

Nanozymes have enzyme-like catalytic activity like that of peroxidase and catalase and finds usage in various field because nanoparticles have the capacity to be modified, functionalized, and mass produced at low cost, and are inherently stable. In this study we used mesoporous palladium@platinum (Pd@Pt) nanoparticles a nanozyme with peroxidase-like catalytic activity that consists of a palladium core with a platinum shell. In addition, these nanoparticles are highly stable with respect to pH and temperature, allow simple conjugation with biological components such as antibodies, and allow simple separation methods through centrifugation. These characteristics allow us to use Pd@Pt nanoparticles as detection agents instead of the enzymes in immunosorbent assays.

Since Collage type II is large enough to have two epitopes, a sandwich immunoassay was used to capture and label this larger analyte with two different antibodies. Assay conditions were optimized for Collage type II quantification using a Pd@Pt nanoparticle-linked immunosorbent assay. An optimal concentration of coating Abs was found to be 2.5 μg/mL at a pH of 8.4 – 8.6, Ab conjugated to Pd@Pt NPs at a level of 7.5 μg Ab to 1 mg/mL Pd@Pt, and a concentration of Ab – nanoparticle conjugates of 2 mg/mL. We plan to apply our sandwich detection method to engineered tissue samples. Our use of this method broadens the scope of Pd@Pt nanoparticle-linked immunosorbent assays beyond our previous study for small analytes such as pesticides. We expect this assay to be adopted for use in the fields of bioengineering, tissue engineering, and bioanalysis.

# 627252 Antimicrobial Surface Coatings to Reduce COVID-19 Spread

**Abstract:**

**Abstract:** The Air Force has identified a critical need to address the COVID-19 pandemic, specifically reducing COVID-19 spread in environments where Airmen operate (e.g., patient transfer mobility aircraft). To address this need, Physical Sciences Inc. (PSI) and National Emerging Infectious Disease Laboratories at Boston University (NEIDL/BU) are developing an antimicrobial coating, and demonstrating its effectiveness on evacuation litter products and blood pressure monitoring equipment. PSI is coating multiple medical equipment materials with a permanently attached, broad-spectrum antimicrobial technology. The coating is being optimized to achieve high levels of viral reduction within a short amount of time. The robustness of coating in litter operation is being evaluated upon weathering, abrasion, and cleaning. We will present the results to date that indicate strong attachment of the antimicrobial coating to surfaces of NATO evacuation litter and blood pressure cuff materials. The coating process was optimized to provide uniform and high-density coverage across all materials. Formulations for both bath and spray coating processes have been developed. All coated materials showed antiviral activity high efficacy (up to 5.5-log reduction, >99.999% kill efficiency) against a COVID-19 surrogate, Vesicular stomatitis virus (VSV). We will also present background information on the coating antimicrobial (AM) activity on textile, metal and plastic surfaces with broad spectrum efficacy against bacteria,+ spores, fungi and viruses. The AM coating demonstrated greater than 99-99.999% kill efficiency against: (a) antibiotic-resistant bacteria: *C. Diff.* (both vegetative cells and spores), *MRSA*, (b) sterilization resistant spores (*Bacillus, sp.*), (c) clean room bacteria (*B. atrophaeus*), (d) Gram-positive bacteria (*S. Aureus, S. Epidermis*), (e) Gram-negative bacteria (*E.Coli*), (f) fungus (*C. Albicans*) and (g) non-enveloped viruses (*MS2*). Biocompatibility of fabric coupons was also demonstrated with no cytotoxicity or skin irritation. We will provide an overview of the production and test methodologies, and discuss the ongoing efforts and next steps.

**Acknowledgement:** This material is based upon work supported by the U.S. Department of Defense under Award AF20R-DCSO1

# 627936 Chemorepellent-Loaded Polymeric Nanocarriers for Biofilm Inhibition


**Abstract:**

Drug-resistant bacterial biofilms cause deadly infections and contaminate industrial food and drinking water systems. Widespread use of antimicrobial treatments has accelerated the emergence of drug-resistant bacteria. As a result, there is an urgent demand for novel technologies to target biofilm formation of specific pathogens.

Our approach exploits chemotaxis, the directed movement of bacteria in a chemical gradient, to control the critical initial step of biofilm formation: bacteria transport to a surface. Inducing chemotaxis requires non-toxic concentrations of chemorepellents, and this is advantageous to reduce selective pressures that cause resistance. However, the competitive dynamics between bacteria transport and chemorepellent release rates for efficient biofilm control are not well characterized. These dynamics were investigated with polymeric nanocarriers (NCs) that allowed for tunable chemorepellent loadings and release rates.

Nickel-loaded polymeric nanocarriers (NCs) were designed to localize a chemorepellent nickel ($Ni^{2+}$) gradient to interfere with transport and subsequent attachment of *Escherichia coli*. NCs were formulated with Flash Nanoprecipitation to encapsulate $Ni^{2+}$ into block co-polymers for polymeric NCs of controlled size and nickel loadings. Dynamic light scattering characterized the nanoparticles to be monodisperse and 100 nm in size. From an ICP-OES $Ni^{2+}$ release study, we found the $Ni^{2+}$ NCs have a rapid and pH-dependent $Ni^{2+}$ release profile, with increased $Ni^{2+}$ release at acidic pH. The $Ni^{2+}$ concentrations released are within the sub-lethal range for *E. coli*.

In our biofilm assay, the $Ni^{2+}$ gradient was localized by coating the $Ni^{2+}$ NCs on a glass-bottom well plate, where *E. coli* HCB1 biofilms were grown and imaged with confocal microscopy. $Ni^{2+}$ NCs significantly reduced the total *E. coli* biomass compared to the control case of NCs without $Ni^{2+}$. Equivalent aqueous $Ni^{2+}$ concentrations did not impact *E. coli* growth. A chemical-in-plug assay with $Ni^{2+}$ NCs revealed that *E. coli* exhibits a chemorepellent response to $Ni^{2+}$ NCs. These assays indicate that rapid $Ni^{2+}$ release from $Ni^{2+}$ NCs induces the non-toxic chemorepellent mechanism that prevents *E. coli* adhesion to the glass surface.

Future work aims to tune NC nickel loading and nickel release rates to develop the required dynamic criteria of chemorepellent treatments for effective biofilm inhibition. As microbes respond to specific chemorepellent compounds, this criteria will be useful to design chemorepellent treatments that target the transport and subsequent biofilm formation of specific pathogens.

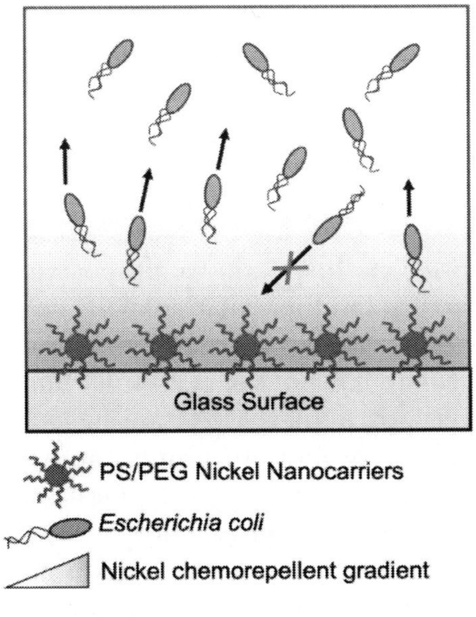
Glass Surface
PS/PEG Nickel Nanocarriers
Escherichia coli
Nickel chemorepellent gradient

# 630774 Semi-Synthetic Outer Membrane Vesicles for Controlled Antibiotic Delivery

**Abstract:**

The rise of antibiotic resistance threatens our ability to treat even minor bacterial infections. With few new drugs in development and rates of resistance increasing, we urgently need new approaches to fight bacterial infections. The goal of this project is to develop a biologically inspired drug delivery vehicle to enable the controlled release of antibiotics. Such a system will permit the use of currently available antibiotics in a more sustainable manner and will overcome certain mechanisms of resistance. Gram-negative bacteria are particularly difficult to treat with antibiotics due to their dual-membrane cell wall, as well as other resistance mechanisms including overexpression of efflux pumps and enzymatic degradation of the drug. In this work, we propose to take advantage of a natural system used by bacteria to deliver biomolecules to other bacterial cells called outer membrane vesicles (OMVs). These OMVs have been demonstrated to deliver functional cargo to the cytosol of Gram-negative bacteria. While this mechanism has not yet been fully uncovered, it suggests the possibility of using OMVs for antibiotic delivery, particularly to Gram-negative bacteria. OMVs provide a number of advantages as a drug delivery system, including stability, long-term in vivo delivery, and the ability to deliver molecules to bacterial cells, including Gram-negative cells. However, they suffer from certain limitations, including loading efficiency and the ability to readily alter membrane properties to improve function ("tunability"). In contrast, liposomes are readily loaded with cargo and their physical properties can be easily tuned with simple adjustments in lipid composition. We therefore hypothesized that a vehicle created from both liposomes and OMVs might facilitate antibiotic delivery and enhance activity in antibiotic-resistant strains of bacteria.

Liposomes of varying lipid composition were mixed with *Escherichia coli* OMVs and extruded to create semi-synthetic OMVs (SS-OMVs). We have demonstrated, using a dye dilution assay, that upon extrusion, the liposomes and OMVs mix thoroughly to create a hybrid vehicle. We verified using dynamic light scattering that the SS-OMVs are stable for at least seven days under a range of storage conditions. In addition, we showed that the SS-OMVs retain their luminal content over this time frame. Using a combination of membrane- and content-mixing assays, we have demonstrated that the SS-OMVs are able to fuse with *Pseudomonas aeruginosa* (PAO1) cells, and current work is focused on optimizing this delivery process. Together, these results demonstrate the potential of a new hybrid antibiotic delivery vehicle for the controlled delivery of antibiotics.

# 632188 Self-Locomotive Antimicrobial Microparticles for Enhanced Biofilm Removal

**Abstract:**

Biofilms are communities of bacterial cells that can cause a variety of healthcare problems. In particular, over 80% of microbial infections in human body is associated with bacterial biofilms, according to a survey from the US National Institutes of Health (NIH). However, traditional antibiotics treatments are not effective enough to remove the wound biofilm because the extracellular polymeric substances (EPS) can impede the diffusion of antibiotics into the biofilm. In this work, we developed a self-locomotive antimicrobial microparticle that can effectively remove biofilms through continuous bubble generation in an antiseptic $H_2O_2$ solution. We hypothesized that the microparticles' continuous bubble generation would deform the EPS matrix and ultimately abrade biofilms due to the repeated bubble expansion and burst. To examine this hypothesis, we firstly improved our previously developed diatom mircrobubbler by increasing the loading mass of manganese oxide ($MnO_2$) via polydopamine (PDA) coating. Then, we assessed and compared their self-locomotive capability by examining the propulsion speed and kinetic reaction in the $H_2O_2$ solution. Using *Pseudomonas aeruginosa* (*P. aeruginosa*) biofilm formed in microgrooves of a silicon substrate, we examined the extent that $MnO_2$ nanocatalyst mass per diatom modulates the abrasion of biofilms. In the end, we evaluated the biofilm removal efficacy by treating *P. aeruginosa* biofilms with minimal biofilm eradication concentration (MBEC) assay.

# 631447 Antibiofilm Activity of Enzyme and pH Responsive Layered Gelatin Nanoparticles

**Abstract:**

Biofilm-associated bacterial infections are one of the greatest threats to health worldwide. Biofilms exhibit a variety of mechanisms that contribute to antimicrobial resistance. Responsive nanoparticles (NPs) that target specific microbial and biofilm features show great potential to prevent and eradicate biofilms. However, developing smart drug delivery systems which respond to multiple bacteria stimuli, thereby increasing specificity to biofilms, remains a challenge. Here, we have developed a hyaluronic acid (HA) and chitosan (CS) coated doxycycline (Doxy) loaded gelatin nanoparticle (GNP) that responds to both bacterial enzymes and changes in pH. Under neutral physiological conditions, the negative charge of the HA coated NPs is expected to enhance blood circulation times when introduced systemically. At biofilm infection sites, bacterial hyaluronidases will degrade the outermost HA layer exposing the underlying CS layer. The NPs will then become positively charged and have greater ability to attach to biofilm bacteria. The acidic biofilm microenvironment will trigger the CS layer to swell providing bacterial gelatinases greater access to the GNP core. Degradation of the GNP will increase Doxy release from the NP, leading to efficient bacteria death and biofilm eradication. Compared to traditional antibiotics, these biopolymer-coated GNPs have the potential to specifically accumulate and deeply penetrate biofilm, which could eradicate bacterial biofilms after controlled release of loaded antibiotics.

By optimizing fabrication conditions (e.g., polymer concentration, assembly time, and pH), GNP, Doxy-GNP, CS-Doxy-GNP and HA-CS-Doxy-GNP were successfully prepared. The average hydrodynamic diameter increased with additional polymer coating layers, from ~215 nm for Doxy-GNP to ~243 nm and ~292 nm for CS-Doxy-GNP and HA-CS-Doxy-GNP, respectively. Additionally, zeta-potential confirmed successful coating, showing charge reversal between uncoated NPs and CS versus HA and CS coated NPs. The encapsulation efficiency and drug loading capacity of HA-CS-Doxy-GNPs was ~25.8% and 6%, respectively. Hyaluronidase treatment led to the most rapid Doxy release kinetics from HA-CS-Doxy-GNP at the conditions tested (150 U/mL hyaluronidase at pH 5), releasing ~40 % of the encapsulated Doxy in 48 hours, compared to only ~18.5 % for NPs in pH 7.4 buffer. These release kinetics were comparable to what was observed when these NPs were incubated in media from a culture of *Vibrio vulnificus*, a highly virulent gram-negative bacterial species. This media was expected to contain hyaluronidases, gelatinases from *V. vulnificus*, with a pH of ~5.5. The biofilm penetration ability of HA-CS-GNP was evaluated against pre-formed *V. vulnificus* biofilms (matured for 48 hours). The eradication ability of HA-CS-Doxy-GNP against *V. vulnificus* biofilm was demonstrated using crystal violet staining, scanning electron microscopy, live/dead staining, and colony counting. Results indicated that HA-CS-Doxy-GNP effectively disperse and eliminate *V. vulnificus* biofilm (nearly 100% elimination of viable bacteria). The biocompatibility of NPs was explored using a cell counting kit-8 assay (CCK-8) assay for human umbilical vein endothelial cells (HUVECs) and NIH 3T3 murine fibroblasts. We found that cells exposed to Doxy loaded NPs for 24 hours displayed lower cytotoxicity than the equivalent amount of free Doxy.

In summary, we have developed a triple stimuli-responsive HA-CS-Doxy-GNP and demonstrated its antibacterial and antibiofilm properties *in vitro*. This bacteria-responsive drug delivery system has the potential to be used to deliver multiple drugs including those aimed specifically at biofilms (e.g., antibiofilm peptides) or signaling molecules for infection detection (e.g., fluorescent dyes) for the effective broad-spectrum treatment and detection of infections.

# 635434 Microbial Growth and Microbial Community Assembly in Household Water Pipes

**Abstract:**

Microbial regrowth occurs in household water pipes where biofilms already established despite potable water being stringently treated, low-nutrient, and supplied with disinfectants. Certain opportunistic pathogens (Legionella spp., non-tuberculous mycobacteria, and Pseudomonas aeruginosa) are enriched in the plumbing environment and are implicated in disease outbreaks. In order to better manage microbial regrowth in pipes, we require better fundamental understandings about the growth kinetics and rules governing the assembly mechanisms of microbial communities. Here, we will discuss microbiome community assemblages in real-world household water from U.S. cities, and the models to explain the regrowth kinetics and community compositions from process and ecological perspectives, respectively. Insights from water microbiome development in drinking water pipes holds relevance to scaling up new materials in contact with real-world fluids.

# 625551 Fighting Bacteria: A Computational Exploration of Bacterial Interactions

**Abstract:**

Tackling issues such as antimicrobial resistance and biomedical material interface interactions requires a deep and fundamental understanding of microbial behaviour. In particular, interactions between microbial communities containing multiple species are complex and not yet well understood, even though they are known to influence microbial survival and disease development. For example, the presence of *Pseudomonas aeruginosa* and *Staphylococcus aureus* in chronic wounds or the airways of individuals with cystic fibrosis has been associated with shortened life spans and decreased lung function. Understanding the movement and interactions of multispecies microbial communities is vital to develop and implement applications relevant to medical and industrial systems, such as infection treatment strategies. This fundamental insight may also provide knowledge required to describe and design systems in which microbes perform a beneficial purpose, such as bioremediation and filtration.

In this work we use computational methods to probe the fundamental interactions of microbial communities. We extend an existing agent-based model (Storck et al., Biophys. J. 106:2037, 2014; van Holthe et al., arxiv, 2020) of self-propelling bacteria to incorporate molecule secreting and sensing as well as polydisperse bacteria shapes. Using this extended model we study the collective dynamics of a bacterial community containing two species. The results are compared to recent experimental work by Limoli et al. (*eLife* 8:e47365, 2019) visualising the early interactions between a multispecies community of bacteria relevant to clinical situations, such as respiratory infections and chronic wounds. The most striking interaction is a 'fighting' process in which *P. aeruginosa*, rod-shaped bacteria, sense secreted products from *S. aureus*, sphere-shaped bacteria, and change their behaviour from collective movement to single-cell exploration with increased velocity. This transition results in a dynamic situation in which the *P. aeruginosa* surround and 'attack' the *S. aureus* groups and significantly inhibit growth of these sphere-shaped bacteria.

This work provides a tool to probe the fundamental controlling mechanisms and resulting spatiotemporal patterns of diverse multispecies microbial communities. From the results, we can learn how we could rationally manipulate interspecies behaviour during infections. Consequently, we might prevent bacteria with harmful effects (such as *P. aeruginosa* and *S. aureus* within the airways of cystic fibrosis patients) from engaging in physical interactions, or bring species together who synergise to produce a beneficial effect.

# 622838 Biofilm Growth Morphology in Confined Heterogenous Media

**Abstract:**

Biofilms, bacteria that produce an extracellular polymeric substance (EPS) that surrounds them, grow in confined structured environments such as the gut, lung mucus, soils, and sediments. The majority of the previous work with biofilms has focused on their behavior on 2D surfaces, such as at liquid-air interfaces or on flat substrates. However, less work has focused on biofilms that grow in confined structured environments. Here, we 3D print *Vibrio cholerae,* a model biofilm former, into jammed hydrogel matrices and observe the growth morphology over time. Through using mutants with varying EPS compositions and controlling the stiffness of the surrounding hydrogel matrix, we develop fundamental insights on how biofilms grow in a confined structured environment.

# 624525 Deconvolute Bacterial Responses to Surface Nanotopography and Surface Chemistry Using Orthogonally Engineered Biointerfaces

**Abstract:**

## Introduction

Reducing bacterial attachment to abiotic surfaces have shown great promise in battling adverse biofilm formation, which has immense importance across various human activities ranging from healthcare to food safety. Nevertheless, despite decades of research and discovery of many successful examples of antibiofouling/antibiofilm surfaces, it is still unclear on a fundamental level how surface properties (e.g., surface topography and chemistry), individually and combinatorially, affect attachment and other bacterial behavior.

One prominent challenge lies in that all surfaces exhibit both topography and chemistry simultaneously to the bacteria they interface with, and that modification of surface topography often inadvertently results in changes in surface chemistry, and vice versa. Furthermore, bacterial cells have been shown to sense and respond to stochastic roughness as small as 100 nm, yet their response to sub-50-nm, well-defined nanotopography is largely unexplored due to low accessibility to sub-50-nm nanofabrication. Finally, surface adherent bacteria undergo dynamic sensing of surface properties, which informs their "decision" to adopt a sessile lifestyle and turn on the biofilm formation sequence – a process underexplored in common biofilm assays. Therefore, the ability to differentiate the decision states of the bacteria during their initial interaction with the surfaces will enrich our understanding of how surface properties affect initial attachment.

## Objective

Here, we aim to *deconvolute* bacterial responses to surface nanotopography and surface chemistry by independently varying these two key surface properties, enabled by orthogonal nanoengineering.

## Methods & Results

*Orthogonal Nanoengineering and Characterization.* First, nanoporous anodic aluminum oxide (AAO) surfaces with highly uniform small (25-nm) and large (100-nm) pores were generated through anodization in oxalic acid. A flat aluminum oxide surface was also prepared as a control. Then, half of these topographically-defined surfaces were further subjected to surface chemistry modification by depositing an ultrathin (< 10 nm) poly(4-vinylpyridine-*co*-divinylbenzene) via initiated chemical vapor deposition (iCVD). The thin film was further rendered zwitterionic (ZW) via exposure to 1,3-propane sultone. As shown in Figure 1A & 1B, the top surfaces and cross-sections of each of the three topographies were well-preserved after surface chemistry modification, and surface chemistry was consistent across surface topographies, as confirmed by FTIR and XPS (data not shown) – hence, orthogonal nanoengineering was achieved.

*Probing Bacterial Responses to Surface Properties.* Next, we introduced a fluorescence-based, quantitative bacterial response assay. A *Pseudomonas aeruginosa* strain, PAO1/pCdrA::*gfp*(ASV), was chosen because it produces a GFP reporter in proportion to the intracellular concentration of cyclic-di-GMP (c-di-GMP), which is a second messenger that regulates bacterial planktonic-to-sessile transition. Thus, more green fluorescence indicates higher c-di-GMP, and probably a cell committed to turning on the pro-biofilm genes. Besides, the PAO1 strain was also stained with DAPI and propidium iodide to reveal the total number of surface-bound cells and cell wall damage, respectively. Together, these fluorescence lables enabled real-time tracking of multiple bacterial responses to the aforementioned surface properties *in situ*, as shown by a proof-of-principle experiment conducted on flat aluminum oxide surfaces (Figure 1C).

Quantitative analysis on such images not only extracted percentage of cells that was alive versus dead, but also provided rich information on the percentage and distribution of cells entering the biofilm sequence, which are likely to become the hotspot for microcolony development. Preliminary experiments conducted on topographically-varied but chemically-identical aluminum oxide surfaces suggest that 25-nm pore surfaces reduce overall attachment whereas 100-nm pore surfaces lead higher cell death (data not shown). Interestingly, both nanostructured surfaces appear to discourage c-di-GMP production. Clearly, these preliminary results require follow-up experiments to confirm.

Ongoing investigations are focused on further isolating the respective effects of surface topography and chemistry individually, as well as understanding the combinatorial effects of the two. Cell tracking with time-lapse images is also underway to investigate the dynamic nature of bacteria-surface interactions as a function of these two surface properties.

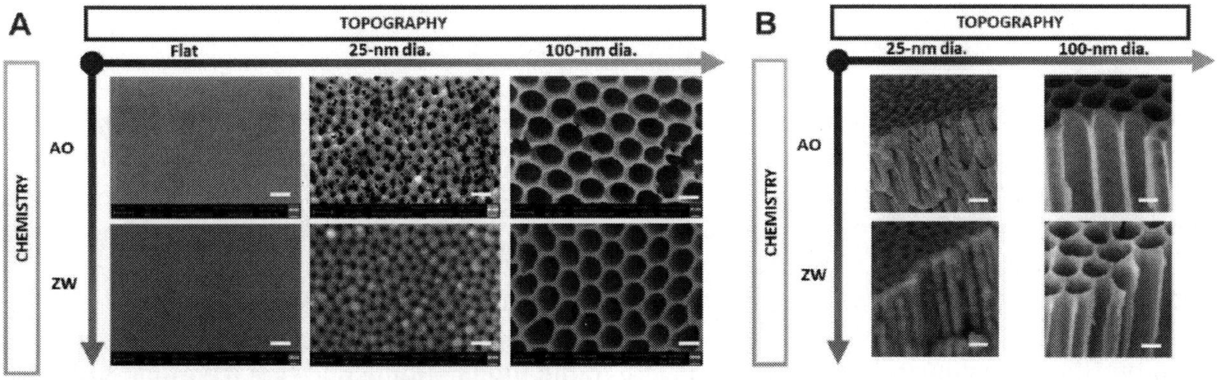


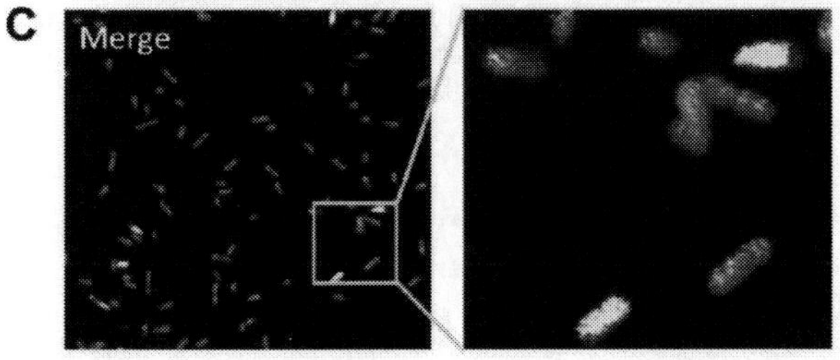


D

| Cell Status | Channel | Percentage |
|---|---|---|
| Dead | PI | 16.7% |
| Live | DAPI – PI | 80.6% |
| C-di-GMP+ | GFP | 2.7% |
| All | DAPI | 100% |

**Figure 1**. Deconvoluting bacterial responses to surface topography and chemistry by using orthogonally engineered biointerfaces. Scanning electron micrographs of the (A) top surfaces and (B) cross-sections of the orthogonally engineered substrates with three different topographies (flat, 25-nm pores, 100-nm pores) and two chemistries (AO = aluminum oxide, ZW = zwitterionic thin coating). (C) Fluorescently stained *Pseudomonas aeruginosa* strain PAO1/pCdrA::*gfp*(ASV) on a flat AO surface, with c-di-GMP production and cell wall damage depicted by green and red fluorescence, respectively. (D) Summary of the different responses

# 631650 Molecular Dynamics Investigation of Biomolecule Adsorption to Graphene and Modified Graphene: Molecular Insights into Biofilm Formation and Adhesion

**Abstract:**

Despite their prevalence in natural and engineered systems (such as biomedical devices), the exact molecular mechanisms and forces controlling biofilm formation and adhesion are relatively unknown. Moreover, surface engineering of these systems to prevent microbial adhesion and biofilm formation has become a key challenge for both the biomedical and materials/corrosion communities. While most modeling studies to date have focused more on continuum or macroscopic film behavior, investigations at the molecular level are needed, too.

Graphene and graphene derivatives are potential candidates to be used as biofilm inhibiting coatings for metal surfaces due to their potential antibacterial properties.[1-2] Chemical vapor deposition (CVD), an alternative to mechanical exfoliation of graphite, is a promising process for large scalable synthesis of graphene. However, CVD usually results in a number of structural defects and surface impurities. While defect-free graphene growth using CVD is still under investigation and these defects remain largely undesired, studies have shown defective graphene surfaces have distinct properties from the pristine layers. What remains to be known is what effect these surface defects have on biofilm adhesion and formation, since no foundational atomistic-level information is available on whether these defects produce positive or negative surface adsorption characteristics relevant to biofilm adhesion and formation.[3]

Recent work has hypothesized that biofilm formation and adhesion may be related to the adsorption of key, early protein molecules including exopolysaccharides (EPS) to surfaces.[4] Nevertheless, studies related to molecular mechanics of these early protein molecule-surface interaction and adsorptive performance of graphene and graphene derivative surfaces is rather scarce. As a first step towards exploring this hypothesis, we investigate the adsorption behaviors of 20 proteinogenic amino acids (as model compounds for microbes), the building blocks of proteins, on pristine and defect-induced graphene surfaces, along with a functionalized surface group (i.e. -COOH, associated with graphene oxide) using molecular dynamics (MD) simulations. All the simulations were conducted in vacuum and in the presence of explicit water as solvent to deduce the effect of solvation on the adsorption behavior.

Specifically, the molecular dynamics simulations are conducted using the LAMMPS molecular dynamics simulation software package and the Adaptive Intermolecular Reactive Empirical Bond Order (AIREBO) and Assisted Model Building with Energy Refinement (AMBER) potentials.[5]–[7] The adsorption energies and the binding free energies of the amino acids on graphene and defective/graphene-derivative/graphene-modified surfaces are evaluated to assess the effect of surface defects on the adsorption phenomena. The binding free energies are computed using the umbrella sampling technique.[8]

The results of this molecular-level study should aid in developing a larger, fundamental understanding of the interaction, adsorption, and adhesion of proteins and microbes to two-dimensional surfaces, such as found in biomedical device applications.

[1] S. V. Agarwalla *et al.*, "Hydrophobicity of graphene as a driving force for inhibiting biofilm

formation of pathogenic bacteria and fungi," *Dent. Mater.*, vol. 35, no. 3, pp. 403–413, 2019.

[2] M. Cacaci, C. Martini, C. Guarino, R. Torelli, F. Bugli, and M. Sanguinetti, "Graphene oxide coatings as tools to prevent microbial biofilm formation on medical device," in *Advances in Experimental Medicine and Biology*, vol. 1282, Springer, 2020, pp. 21–35.

[3] S. P. Singh, S. Ramanan, Y. Kaufman, and C. J. Arnusch, "Laser-Induced Graphene Biofilm Inhibition: Texture Does Matter," *ACS Appl. Nano Mater.*, vol. 1, no. 4, pp. 1713–1720, 2018.

[4] J. Wang, K. M. Goh, D. R. Salem, and R. K. Sani, "Genome analysis of a thermophilic exopolysaccharide-producing bacterium - Geobacillus sp. WSUCF1," *Sci. Rep.*, vol. 9, no. 1, pp. 1–12, 2019.

[5] S. Plimpton, "Short-Range Molecular Dynamics," *J. Comput. Phys.*, vol. 117, no. 6, pp. 1–42, 1997.

[6] E. Darian and P. M. Gannett, "Application of molecular dynamics simulations to spin-labeled oligonucleotides," *J. Biomol. Struct. Dyn.*, vol. 22, no. 5, pp. 579–593, 2005.

[7] S. J. Stuart, A. B. Tutein, and J. A. Harrison, "A reactive potential for hydrocarbons with intermolecular interactions," *J. Chem. Phys.*, vol. 112, no. 14, pp. 6472–6486, 2000.

[8] G. M. Torrie and J. P. Valleau, "Nonphysical Sampling Distributions in Monte Carlo Free-Energy Estimation: Umbrella Sampling," *J. Comput. Phys.*, vol. 23, p. 187, 1977.

# 636265 Invited Talk: Machine Biology for Infectious Diseases

**Abstract:**

Machines have the potential to outperform humans and revolutionize our world. In this talk, I will describe our efforts using machines to develop computational approaches for antibiotic discovery, as well as low-cost rapid diagnostics. Computers can already be programmed for superhuman pattern recognition of images and text. In order for machines to discover novel antibiotics, they have to first be trained to sort through the many characteristics of molecules and determine which properties should be retained, suppressed, or enhanced to optimize antimicrobial activity. Said differently, machines need to be able to understand, read, write, and eventually create new molecules. I will discuss how we trained a computer to execute a fitness function following a Darwinian algorithm of evolution to select for molecular structures that interact with bacterial membranes, yielding the first artificial antimicrobials that kill bacteria both *in vitro*and in relevant animal models. My lab has also developed pattern recognition algorithms to mine the human proteome, identifying throughout the body thousands of antibiotics encoded in proteins with unrelated biological function, and has applied computational tools to successfully reprogram venoms into novel antimicrobials. I will also describe the development of low-cost diagnostic biosensors for COVID-19, further substantiating the exciting potential of machine biology. Computer-generated designs and innovations at the intersection between machines and biology may help to replenish our arsenal of effective drugs and generate novel diagnostics, providing much-needed solutions to global health problems caused by infectious diseases.

# 636263 Invited Talk: Nucleic Acid Mimics-Based Strategies to Tackle Antimicrobial Resistance in Bacteria

**Abstract:**

Nucleic acid mimics (NAMs) or nucleic acid analogues can be engineered to hybridize with specific sequences in bacteria with high affinity and can therefore be used as antisense antibacterials, inhibiting the translation of essential genes in bacteria. However, as the bacterial envelope barrier hinders NAMs internalization, suitable delivery vectors are needed. In here, I will discuss limitations and advantages on the application of liposomes, dendrimers, vitamin B12 and cell-penetrating peptides as carriers for NAMs. Important aspects to discuss include a stable association between the NAMs and the carriers in the presence of biofluids, the type of interaction between the carriers and the cell envelope and also the accumulation of NAMs inside the cell that each carrier is able to achieve.

# 636193 Invited Talk: The Biofilm Matrix As a Therapeutic Target

**Abstract:**

Biofilm-associated chronic infections are notoriously recalcitrant and exceedingly tolerant to antibiotic treatments. It is thought that the biofilm matrix, a hydrated layer or polysaccharides, proteins and nucleic acids that surrounds microbes in a biofilm, is a primary driver of chronicity and tolerance during infection. We hypothesized that by degrading these matrix components, we can expose the microbes inside the biofilm and increase antibiotic efficacy. Here I will discuss our pre-clinical evaluation of polysaccharide-degrading enzymes for wound infections, and what we have learned about the biofilm matrix of *Pseudomonas aeruginosa* during infection.

# 636117 Invited Talk: Probing Bacterial Social Behaviors during Human Infection

**Abstract:**

The survival of pathogens in the human body has been rigorously studied for well over a century. Bacteria are able to colonize, persist and thrive *in vivo* due to an array of capabilities, including the ability to attach to host tissues, produce extracellular virulence factors, and evade the immune system. Most bacterial pathogenesis studies have focused on mono-culture infections; however, it is clear that many bacterial infections are not simply the result of colonization with a single species, but rather ensue from the actions of polymicrobial communities. Here, I will discuss how interactions between bacteria impact community development, resistance to host innate immunity, and persistence in experimental model and human infections. I will also discuss the use of novel technologies including micro-3D printing and scanning electrochemical microscopy for probing bacterial interactions in biofilm populations.

# 636264 Invited Talk: Harnessing Synthetic Biology to Characterize and Treat the Gut Microbiome

**Abstract:**

Synthetic biology is bringing together engineers, physicists and biologists to model, design and construct biological circuits out of proteins, genes and other bits of DNA, and to use these circuits to rewire and reprogram organisms. These re-engineered organisms are going to change our lives in the coming years, leading to cheaper drugs, rapid diagnostic tests, and synthetic probiotics to treat infections and a range of complex diseases. In this talk, we highlight recent efforts to create living therapeutics and diagnostics as well as paper-based diagnostics for characterizing and treating the guy microbiome.

# 629865 Rethinking Convective Chromatographic Separations

**Abstract:**

The use of membrane supports with immobilized protein A has not lead to a technological breakthrough as it was expected when the first protein A affinity membranes appeared on the market in the late 80's/early 90's. Their excellent flow properties were not sufficient to compensate the low binding capacity with respect to chromatographic beads for protein capture applications. Nevertheless, the development and booming of single use technology explains the success of membrane adsorbers for impurity removal in biopharmaceutical manufacturing.

Besides the widespread application of membrane chromatography in polishing steps, there is still a large and growing research interest in its implementation for capture and purification of high added value products such as antibodies and other therapeutic molecules. This research effort led to a remarkable progress in developing highly selective and efficient stationary phases. Indeed, the binding capacity of protein A immobilized on an optimized membrane base material, with respect to commercially available protein A membranes, could be increased about seven-fold that demonstrates the tremendous effect that the membrane support has on binding capacity. However, these values are still too low with respect of the capacity of chromatography beads, even if competitive binding capacities have been obtained with nonwoven ion-exchange membrane adsorbers.

Moreover, continuous biomanufacturing will underline the limits of the conventional chromatographic process based on packed bed columns and highlight the need of a novel chromatographic process. In this work, innovative supports for convective chromatography produced with 3D printing will be presented. These new stationary phases, inspired by nature, have the potential to dramatically change chromatographic separations.

# 632551 Covalent Organic Framework As an Ion Exchange Material for Adsorptive Separation of Biomolecules

**Abstract:**

In this work, the use of covalent organic frameworks (COFs) in biomolecule separation has been explored. We examined the ionic COF Py-BPy$^{2+}$-COF as an ion exchange material for protein and amino acid separation. After characterizing the physical and chemical properties of the synthesized COF with a variety of techniques such as Fourier transform infrared analysis, powder X-ray diffraction, X-ray photoelectron spectroscopy, and transmission electron microscopy, binding experiments with both large and small biomolecules were performed. High adsorption capacities of amino acids with different hydrophobicity and charge, as well as proteins of different isoelectric points and molecular weights were determined in batch equilibrium experiments. Desorption experiments with model proteins mixture demonstrated an ability to successfully separate one protein from another with selectivity hypothesized to be a combination of isoelectric point, hydrophobicity, and ability to penetrate the crystalline material. The results demonstrated that Py-BPy$^{2+}$-COF can be exploited as a robust crystalline anion exchange biomolecule separation material.

# 632787 Scalable Synthesis of Nanoporous Atomically Thin Graphene Membranes for Dialysis and Molecular Separations Via Facile Iso-Propanol-Assisted Hot Lamination

**Abstract:**

Scalable graphene synthesis and facile large-area membrane fabrication are imperative to advancenanoporous atomically thin membranes (NATMs) for molecular separations. Although chemical vapordeposition (CVD) allows for roll-to-roll high-quality monolayer graphene synthesis, facile transfer withatomically clean interfaces to porous supports for large-area NATM fabrication remains extremely challenging.Sacrificial polymer scaffolds commonly used for graphene transfer typically leave polymer residues detrimentalto membrane performance and transfers without polymer scaffolds suffer from low yield resulting in high non-selective leakage through NATMs. Here, we systematically study the factors influencing graphene NATMfabrication and report on a novel roll-to-roll manufacturing compatible isopropanol-assisted hot lamination (IHL)process that enables scalable, facile and clean transfer of CVD graphene on to polycarbonate track etched(PCTE) supports with coverage ≥99.2%, without compromising support integrity/porosity. We demonstrate fullyfunctional centimeter-scale graphene NATMs that show ~2-3 orders of magnitude higher permeance andbetter selectivity than commercially available state-of-the-art polymeric dialysis membranes, specifically in the0-1000 Da range. Our work highlights a scalable approach to fabricate graphene NATMs for practicalapplications and is fully compatible with roll-to-roll manufacturing processes.

Cheng et al. Nanoscale. 2021
Cheng et al. Nano Lett. 2020
Kidambi et al. Adv. Mat. 2018
Kidambi et al. Adv. Mat. 2018
Kidambi et al. ACS App. Mat. Int. 2018
Kidambi et al. Adv. Mat. 2017
Kidambi et al. Adv. Mat. 2017
Kidambi et al. Nanoscale 2017

# 627558 Immiscible Liquid-Coated Materials for Bioseparations

**Abstract:**

Bioseparations, such as liquid and aerosol filtration, are most often performed with the use of porous materials. However, porous materials can be prone to biofouling and other types of surface adsorption. In this work, we present bio-inspired liquid-coated surfaces as a new approach to separation materials that resist fouling. Liquid-coated filters can be created by immobilizing a water immiscible liquid on the surface of commercially available synthetic filtration membranes. By chemically matching the infusing liquid to the filtration membrane, the stable liquid layer that is formed produces a dynamic, mobile surface that resists the adhesion of contaminants. For water purification, 0.45 μm pore diameter polytetrafluoroethylene (PTFE) and polyvinylidene fluoride (PVDF) filters were tested using omniphobic perfluoropolyether coating liquids. We measured the continuity of the surface liquid layer by testing how easily a water droplet could begin to move the surface, as well as the anti-adhesion properties of the surface liquid layer through quantifying the speed of droplet movement at a static angle of inclination. The results indicate that the PTFE membranes could sustain a more consistent functional liquid layer, with an approximately 75% lower sliding angle and 70% faster droplet movement at a fixed angle than the PVDF. For aerosol filtration, we tested liquid-coated commercial PTFE and HEPA filters for their efficiency in capturing and releasing aerosolized *Escherichia coli* for analysis. We determined that bacteria trapped on the liquid-coated filters could be removed with increased efficiency compared to bare controls. The use of liquid coated materials in bioseparations opens new doors for the creation of dynamic filtration strategies.

# 623769 Cu-Selective Membrane Adsorbers for Medical Isotope Production


**Abstract:**

Radiopharmaceuticals offer promising new approaches for imaging and treatment for cancer patients; however, the current supply of these novel drugs is too low to support clinical trials. Copper-67 (Cu-67), can provide treatment (via beta decay) and imaging (via gamma rays). Cu-67 is purified using resin-packed columns, which are known to suffer from diffusion-limited transport. The practical implications of these limitations are (1) low column flow rates, (2) high elution volumes and (3) long purification times—all of which contribute to the bottleneck in the sustained supply of Cu-67. Due to the short half-life (2.58 days) of Cu-67, rapid purification is key for increasing the availability of the isotope for clinic trials and use.

A promising alternative to resin-packed columns are membrane adsorbers, which have seen recent success in the purification of biologics. This contribution describes our efforts to graft glycidyl methacrylate (GMA) from polyvinylidene fluoride (PVDF) membranes through activator generated by electron transfer, a type of, atom transfer radical polymerization (AGET ATRP). After grafting poly(GMA), diamine ligands of varying chain length (putrescine and ethylene diamine) are attached to the brushes through an epoxide ring opening reaction. Membranes are characterized by attenuated total reflection Fourier-Transform infrared spectroscopy throughout the synthesis process (grafting and ring-opening) to support functionalization. Membrane permeability was calculated from pure-water flux experiments in a dead-end filtration cell to be 5166 LMH/bar for unmodified PVDF membranes, 4692-648 LMH/bar for GMA modified membranes, and 4124-175 LMH/bar for the amine functionalized GMA membranes. Membrane morphology and surface chemistry was examined using electron microscopy (SEM) and SEM x-ray photoelectron spectroscopy (XPS). XPS analysis confirms that the Cu catalyst was successfully washed from the membrane after AGET ATRP. Batch adsorption experiments were conducted for Cu and competitor ions (Ni, Zn) in pH 2 HCl solutions. Ion concentrations were measured using an inductively-coupled plasma optical emission spectrophotometer (ICP-OES) and the binding capacity was calculated by modeling equilibrium adsorption data with the Langmuir isotherm. The results of this research are laying the groundwork for implementing membrane absorbers as a new separation material for medical isotope purification.

# 628735 Tuning Structural Defects on Nominal Single-Layered Graphene Oxide Membrane for Selective Separation of Biomolecules

**Abstract:**

The advent of 2-dimensional materials provided a unique opportunity to design and synthesize the ideal membrane with ultrathin thickness and precisely controlled molecular-sized pores. Graphene oxide (GO), owing to its water solubility and functionality, has attracted enormous attention for membrane applications. However, standard synthesis of an ideal high-performance membrane is still limited by the thickness of GO membranes, leading to a loss in permeance. Optimizing the deposition methodology of GO flakes on polymeric substrates could address this limitation and enable the synthesis of ultrathin single-layered membranes. In this study, a sequential coating methodology was adopted for depositing single layer GO flakes on a Polycarbonate polymeric support with increments of 10% area-based coating of GO, followed by different time exposure to oxygen plasma in order to form tunable transport pathways on the surface of GO. Also, retention of different single model proteins (BSA, Lysozyme & IgG) and of mixtures of these proteins were measured. In addition, the separation efficiency for similar sized proteins (myoglobin and lysozyme) exhibited separation factors of ~6 with purities of ~92% through the plasma-etched GO membrane. These findings offer new opportunities of using GO flakes to obtain ultrathin membranes on polymeric supports and fabricate tunable GO membranes for possible application in biotechnology industry.

# 626234 The Development of Nanocarbon Immobilized Membrane for Elimination of Thermophilic Bacteria Via Membrane Distillation

**Abstract:**

Pathogen contamination in water has always been a threat to public health causing diseases such as dysentery, typhoid and cholera and such outbreaks have led to millions of deaths worldwide. Methods for treating pathogen-contaminated water include boiling, treatment with ozone, chlorine, UV irradiation, and filtration. These methods have their limitations, can be expensive, and can also produce secondary hazardous wastes such as disinfection byproducts that themselves are toxic. Recently, membrane processes which have low cost, smaller footprint, and provide effective treatment options for a variety of pollutants have become popular in water purification.

Membrane distillation (MD) is a separation technique where the driving force is a vapor pressure gradient between a hot feed and a cold permeate. While many microorganisms get killed at higher temperatures, the challenge arises in cases where these microorganisms are stable at high temperatures namely thermophilic bacteria. In this study, carbon nanotube, functionalized carbon nanotube, and graphene oxide immobilized membranes were used in membrane distillation to generate microorganism free water. The nanocarbons degraded the bacterial cell integrity and caused cell death. The pure CNTs showed the highest biocidal activity (96.2%) followed by functionalized CNTs and graphene oxide. The biocidal activity was significantly higher than the plain polymeric membranes.

In general, the biocidal action of nanocarbons typically involves a combination of physical and chemical mechanisms. First, the nanostructured CNTs acted as a mesh to trap the bacterial cells. The presence of van der waals forces facilitated the adhesion of bacterial cells to the surface of CNTs. The sharp structures of CNTs are also known to cause significant structural damage to the cell wall and the membrane of the microorganism. On the other hand, GO exists as layers of nanosheets, and we believe that initially the bacterial cells get trapped within the nanosheets. Bacterial cell death in the case of GO is believed to be caused via oxidative stress. The presence of GO induces the production of reactive oxygen species (ROS) within the bacterial cells, which causes a decrease in membrane fluidity by altering membrane properties, ultimately leading to cell death. The CNT membranes also showed less fouling. While biofilm/bacterial sludge accumulated on the CNT-COOH and GO immobilized membrane surfaces, no biofilm was observed on the CNT membrane. Among the membranes studied, the CNTs appeared to have the highest biocidal effect and lowest biofouling.

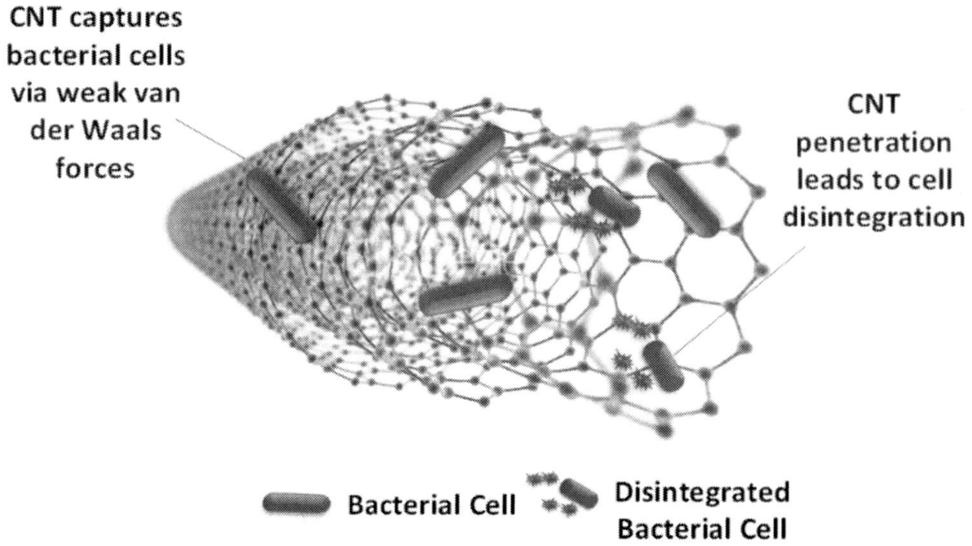
CNT captures bacterial cells via weak van der Waals forces
CNT penetration leads to cell disintegration
Bacterial Cell
Disintegrated Bacterial Cell

# AUTHOR INDEX